Jane Goodall at 90

ISBN 978-1-62806-392-9 (print | paperback)
ISBN 978-1-62806-393-6 (ebook)

Library of Congress Control Number 2023922161

Published by Salt Water Media
29 Broad Street, Suite 104
Berlin, Maryland 21811
www.saltwatermedia.com

Front cover image of Jane and Loup, Tom's dog, in 2005 having fun in Tom's old red pick-up truck, a 1949 Studebaker used courtesy of Thomas D. Mangelsen; back cover image of Jane smiling in the sun used courtesy of Thomas D. Mangelsen — mangelsen.com

Cover design by Salt Water Media

Interior images are provided by and used courtesy of the image source noted in the caption and the small chimpanzee image is a stock image used with license

Jane Goodall at 90

Celebrating an Astonishing Lifetime of Science, Advocacy, Humanitarianism, Hope, and Peace

Edited by Marc Bekoff and Koen Margodt

Contents

A Loving Tribute to a Most Amazing, Iconic, and Indefatigable Woman—and Much More

The essays in this book come from every habitable time zone in the world and every continent except for Antarctica. Neither of us could ever have imagined that Jane, herself, could ever have imagined, that she would evolve into being, and over the years becoming, the global icon she is today. As Marc notes below, when he spent time with Hugo van Lawick in the early 1970s and heard about all of the amazing things Jane was doing way back then way off in what was then called Tanganyika he too, could never have foreseen the huge influence she has had over people in countless countries and in his own work.

Jane taking a "selfie" during her early years in Gombe. Jane put the camera in a tree.

Image Source: Jane Goodall

On April 3, 2024, Jane is 90 years old! We both could be shocked having worked with her in the recent past as she globe-trotted here and there and tirelessly gave lively lectures and gave everyone who wanted her attention what they were looking for and needed.

However, we're not all that surprised. While we are in awe of who Jane has become with unrelenting hard work and an everlasting umbrella of hope hanging over her, we're both thrilled to offer this book as a huge thank you to our friend and esteemed colleague, Dr. Jane Goodall, DBE, United Nations Messenger of Peace, and founder of the Jane Goodall Institute (JGI) and Roots & Shoots (R&S).

We're at a loss for words for how to describe Dr. Jane, as she is often called. Both of us have known Jane for many years and have worked closely with her on a number of projects during which we had serious discussions, fun, laughter, and ample amounts of peaty scotch. We also have co-chaired JGI's Global Ethics Committee for several years and have benefitted from Jane's countless words of wisdom.

This book is a follow up to *The Jane Effect: Celebrating Jane Goodall* that was presented to Jane – hand-delivered to Jane by Tom Mangelsen – on her 80th birthday, April 3, 2014! We see this unique collection of essays as somewhat more "personal" than the previous collection of accolades and there was no way we could include everyone – the many thousands of people – on

Jane and Marc playing tug-of-war before a talk in Denver, Colorado in March 2023.

Image Source: Marlon Reis

whom Jane has had a significant positive effect. So, we hope readers will sit back and learn more about Jane, and we hope – and know – there will be many surprises awaiting.

Our original idea was to have people write about Jane as an advocate, globe-trotting voice for animals, humans, and their homes, a pillar of hope, and scientist. However, as essays came in, we realized that it was nearly impossible to pigeon-hole Jane into this or that category so, for convenience, we decided to organize these heartfelt contributions alphabetically by the first name of the authors. It's immediately clear that Jane can easily be called a "Jane-of-all trades" – a tireless scotch-drinking, globe-trotting, eclectic English woman – that is how diverse she truly is. She also is an academic with a strong practical on-the-ground bent, nimble, athletic, wonderful storyteller, and much more.

We are so pleased to surprise you with this book as a birthday gift. It brings together a wide variety of essays and views into your amazing life ranging from family, close friends, primatologists, zoologists, anthropologists, members of the JGI family, animal advocates, conservationists, actors, spiritual leaders, political leaders, and entrepreneurs from around the world. You are a magnet and easily attract diverse people of all ages and cultures into your global umbrella of caring, compassion, empathy, dignity, wisdom, and hope.

Some express brief sentiments, others explain how you inspired them, what drives them to do what they do, while others simply share some fun and joyful memories. We didn't want to impose a strict script and we are quite confident that you prefer this freedom yourself! We wanted this to be a celebration, informal and fun, and to let every contributor express themselves in their own unique way and voice!

Although we've known you for many years, we were enthralled to learn things we have never previously heard. We hope you will enjoy these memories as much as we did as we put this book together, and we're sure readers also will. In many ways this book brings a picture of Jane and those accompanying you on this wonderful journey that hasn't been told *anywhere else*. Time and time again we were receiving essays which gave us goosebumps. There are testimonies on how your first scientific lecture was received back in 1962, how you were continuing your message of hope in the Democratic

Jane "bathing" and enjoying a glass of wine
as Mr. H. carefully looks on in Nebraska, 2013.

Image Source: Thomas D. Mangelsen

Republic of Congo although shelling became louder and louder, your care for Jezebel the spider, the challenges your translators faced and many other treasures we never heard of or read about before. Treasures about you, but also about the wonderful, heartwarming work so many other people undertake. It is dazzling to see how deeply you inspire so many of us into doing good. How many children, teenagers and adults are inspired by you? How many humans and animals have been helped by you and those inspired by you? How much habitat has been saved, how many trees planted?

We see this book as a treasury of memories. One could also call it a birthday cake with ninety virtual candles, ninety tiny flames of hope. We know you enjoy pictures and we've also brought a selection of visual memories together, as cherries on your cake. There's a few you may be seeing for the very first time!

We've been secretly celebrating you and your birthday for nearly a year. We had tons of fun working with one another and connecting with so many ambassadors of hope around the world. We hope you'll enjoy this book as much as we did and that other readers will do so as well.

The most challenging part of this project was to choose who to invite to write about your amazing life. There are simply too many people who love you! As Mary Lewis mentioned, we could easily compile an encyclopedia with people who would eagerly be willing to contribute. We apologize and hope nobody feels intentionally left out—it was difficult to locate some people and some likely don't check their email very frequently. Nonetheless, it's heart-warming to see how many people are following your lead globally. Their stories turn this little book in an energy booster, inspiring us to act every day for a better, brighter, and more compassionate world. You know them. You know what they do. But we hope you'll still enjoy these brief testimonies, these little flames shining on your wonderful birthday cake.

Thank you for keeping our and other people's hope, hearts, and dreams alive. Thank you so, so much for all you're doing!

With much love and affection,

Marc and Koen

Some Words from Marc Bekoff

Although Koen and I have worked together with Jane on various projects over many years, we have different histories with her and I, as a world-renowned ethologist, behavioral ecologist, and champion of compassionate conservation, have this to say:

Back in the early 1970s when I was a graduate student at Washington University in St. Louis, Missouri, I'd heard about Jane Goodall going off to live with the chimpanzees of Gombe. I also knew of her husband at the time, the renowned National Geographic photographer Hugo van Lawick. I'd already read Jane's wonderful and groundbreaking monograph on the behavior of free-living chimpanzees in the Gombe Stream Reserve, and it was clear that she was well on the way to making a difference in how animals were studied and the ways in which people referred to them. I pocketed those perspectives, and they have always been in my head and heart in my own studies over the past four-plus decades.

A fax from Jane to Marc
and his dog, Jethro,
as Jane and Marc were writing
The Ten Trusts.

Image Source: Marc Bekoff

In fall of 1971, when I was still in graduate school, an unexpected visitor showed up at my home in St. Louis after letting me know he was traveling around the country. It was Hugo. He stayed there for a few days and shared his bed with Moses, a huge white malamute. Hugo and I had long chats about animal behavior, the importance of observing identified individuals over long periods of time, and what Jane was accomplishing – despite a large number of skeptics. It's well-known that Jane's seminal observations of David Greybeard using a tool were met with skepticism until she showed a video of this amazing behavior.

Yes, that was an exciting discovery, but Jane's methods and approaches to animal behavior were what I really found so astonishing – starting with her habit of naming the awesome chimpanzees she studied and stressing their individual personalities. She always felt that every individual counts, not only among the animals she was studying but also when working with people who were concerned about saving other species and their homes. At the time, naming and recognizing individuality were not standard operating procedure in studies of animal behavior, most of which were conducted in artificial situations in various sorts of captive settings. "Naming animals is too subjective and it'll influence how data are explained," I was told, and individual differences were "noise in the system," while talking about animal personalities was fraught with error. At the time, most researchers engaged in normative thinking about other animals and liked to talk about "the dog," "the coyote," "the chimpanzee," or "the elephant." Subsequent research has shown how wrong they were. My Ph.D. mentor, Michael Fox, fully supported what Jane was doing. He also fully supported me when colleagues in our department said I couldn't name the animals I was studying nor should I be talking about personalities. How unscientific it was, they said: Animals should be numbered and only humans had personalities.

I was vulnerable, of course, as a mere graduate student, but I had Jane's example to support me. I knew that Jane had refused to change the ways in which she referred to the chimpanzees, and I too refused to change. In the end, it worked. And, over the past forty some-odd years, Jane has been proven to be right on the mark. Science has changed, and we are now allowed to consider animals as subjects, not objects, and to recognize that their individual personalities are extremely important to study.

I met Jane on a few occasions during the 1970s and 1980s, and in 1999 we got together when she was in Boulder, staying at the home of a mutual friend. We hit it off, wrote a few essays together, and then worked together on a book that was published in 2002 called *The Ten Trusts: What We Must Do to Care For the Animals We Love,* that has been published in a number of foreign languages. Working with Jane on that book and on other pieces was a true joy, and despite her horrific travel schedule she always was there for talking and faxing. Jane has this uncanny ability to give her full attention to someone despite being pulled here and there. Jane and I also cofounded Ethologists for the Ethical Treatment of Animals: Citizens for Responsible Animal Behavior Studies in July 2000.

When we began working together Jane wasn't using email. How could that be? What a pain it was. We often crossed paths through faxing or phoning, and when I awoke and began working at around 4 in the morning, Jane was eating her typical small lunch or having tea with her sister, Judy, and other friends. I also began to work closely with Jane's Roots & Shoots program as a roving ambassador, because I too travel all over the world – though not as much as Jane, of course. I work with youngsters, senior citizens, and inmates. Many of the students in my class at the Boulder County Jail are exceptional writers and artists, and their creative activities bring them a lot of joy and hope.

This is a drawing of Fifi, one of Jane's favorite chimpanzees, by Geoff, a student in my class on animal behavior and conservation at the Boulder County jail. It won an award at an art context to which I submitted it.

How Jane continues to do what she does blows my mind, and she is still going strong as she turns 90 years old. I share fully Jane's belief that every individual counts and that everyone can make a positive difference in the lives of other animals and in saving their homes.

Jane clearly is one of the most influential scientists and spokespersons for animals *in history*. She also has been a tireless advocate for humans. Sometimes I hear people say she really isn't a scientist, and how wrong they are. Her original monograph is a classic, as is her later monograph – the

Jane and Marc working on one of Jane's books
at Tom Mangelsen's cabin in Nebraska in March 2012.

Image Source: Thomas D. Mangelsen

encyclopedic tome called *The Chimpanzees of Gombe* published in 1986 summarizing much (though by no means all) of what she had learned about chimpanzees in her first twenty to twenty-five years of research. Jane also works closely with human animals because she understands there's no way to work for nonhumans without figuring out how humans can peacefully coexist with them.

On the personal side of things, Jane and I share a passion for good single malt scotch, and when we meet here and there, I always bring a small flask of what she calls "cough medicine."

Suffice it to say, Dr. Jane has influenced my life in many, many ways, and I am thrilled to count her as a close friend and to contribute to celebrating her 90th birthday. Thank you, Jane, for what you have done, and are continuing to do, for all animals, nonhuman and human, and for their homes. I carry you and your messages in my heart and will continue to do so forever.

Lots of love,
Marc

Jane's Special Cough Medicine
Image Source: Marc Bekoff

Jane and Marc toasting at Tom Mangelsen's cabin
in Nebraska during the crane migration, March 2012.
Image Source: Thomas D. Mangelsen

Some Words from Koen Margodt

Jane and I met some thirty years ago. I had just finished my studies in moral philosophy, where I had a special interest in animal ethics. It was written in the stars – as a child I already had a strong passion for animals. In elementary school, I'd always take the greyhound Tory for a walk. Tory lived on the neighbouring farm, where he was locked most of the day in a dark stable. We both loved these walks. Later, while writing my master's thesis on zoo ethics, it struck me how chimpanzees were living in very grim circumstances in substandard zoos. But there's another reason, I made contact with you. You were open to talking with people and reaching out to those you disagree with to have a decent conversation. That strongly appealed to me.

I was a pretty shy person, but the fate of the chimpanzees encouraged me to act. Those were days without internet. I learned from your book *Through a Window* that your mother, Vanne, lived in Bournemouth. I got her number and called her to obtain your postal address in Tanzania. To my utter surprise, you picked up the phone, I stumbled through our conversation and a few months later we were talking in Brussels. We worked together to send letters to the directors of those substandard zoos and we made a difference in the lives of several chimpanzees. When I visited a zoo director near Ghent, Belgium, he promised to get the chimp Toto out of his dark enclosure and to build a new one with trees. And he did and introduced Toto to another chimpanzee coming from a circus. Like for so many other animals, you were the changemaker.

All this was the start of a wonderful collaboration and many other things followed. Helping chimpanzees around the world, lectures in Belgium, setting up JGI Belgium with others, raising funds for Tchimpounga, planting trees, and encouraging Roots & Shoots groups. You kindly agreed to be an advisor for my Ph.D. thesis on the moral status of great apes. Over the last years we've been working closely together on a variety of topics involving

Jane and Koen after a JGI Belguim Board meeting in 2016.
Image Source: Koen Margodt

chimpanzees, pangolins, rats, dogs, dolphins, and many other animals. And you always give yourself 200%, with a sense of urgency, positive energy, and passion. "Koen, this isn't our duty, it's our passion."

Your journey is not one about fame, praise, idolatry, or money. It's about caring for others, influencing others, and making a positive change. The enthusiasm of young Jane who set up the Alligator Society during the 1940s has always remained. Sometimes I feel like you've simply expanded this 'very first Roots & Shoots group' around the globe!

And there's also caring for humans. You're not merely observing as an ethologist, you connect and care. You're aware of everyone in the room and easily build relationships with all – whether they enter the spotlight or – especially – those remaining in the shadows of the back row, whether they are humans or animals, young or old.

Among all topics on which we collaborated, there's always been attention for the personal side. You may go silent for weeks, but if there's a health issue in the family, you know about it. And you're there to listen or bring consolation. Even months later you may, out of the blue, suddenly check

A selfie by Mana Margodt with Jane and Koen during a fundraising dinner in Belgium, 2022.

Image Source: Mana Margodt

in to see how things are going. I've no idea how you keep track – it must be your big heart.

People often ask how you keep going. There's definitely your positive spirit, a sincere belief that every individual matters and can make a difference, one step at a time. Moreover, you have a wonderful sense of humor, and you tell jokes. I made a mistake only a few years ago when I asked whether I should give you an arm while descending a long stairway of concrete towards a lecture hall. You instantly took off, hopping and dancing downstairs while I was holding my breath, then waiting for me with a mischievous smile. Nobody tells Jane what to do!

In 2022 there was this beautiful, ancient wooden stairway at a JGI meeting in Vienna. Staff wouldn't allow us taking a group picture with JGI friends, although the staircase was solid wood. You asked me what the fuss was about. Before I had finished explaining, you were already upstairs waving to all of us, raising your glass to all of us!

Oh, and lastly – a very happy birthday and thank you for everything! May your young heart keep on bouncing for many more years. Reading these essays astonishes me even more about how many people you've inspired to act for a better world and how much impact we make together. Working

Mary Lewis, Koen, and Jane in Vienna, 2022.

Image Source: Koen Margodt

together with friends inside and outside JGI is a true energy booster and lots of fun! There are feathers of Jane bird around the world, an energy that will keep on soaring in the sky and never fading away. The younger generations are stepping forward and I'm proud of our daughters Mana and Fara who are undertaking, inspired by you, a variety of actions, such as cleaning up our neighbourhood, helping children from less affluent communities with their homework, raising funds for a little school in Tanzania or volunteering in a children's hospital in Nepal. As you said, standing recently on a chair during a fundraising dinner in Brussels – "Together we can, together we will!"

Much love,
Koen

Jane on the ancient wooden stairway in Vienna, 2022.

Image Source: Koen Margodt

JGI Friends in Vienna, 2022.

From left to right:
Itai Roffman (JGI Israel)
Koen Margodt
Steve Woodruff (JGI Global)
Aslihan Niksarli (R&S Turkey)
Jane Goodall
Bella Lam (JGI Canada)
and Galitt Kenan (JGI France)

Image Source: Koen Margodt

Giving Thanks to All

Organizing this book has been a labor of true love and learning how much love so many other people have for Jane and all she's done and continues to do. We are so grateful to all contributors living all over the world for sometimes digging deeply into their memories and sharing their experiences. We've been chasing you with our mails and reminding you about deadlines, and we are ever so grateful for our collaboration. We understand that for some contributors this project required quite a considerable effort, whether due to workload or health. Thank you for joining us on this journey. We hope you will be as proud about this book as we are.

We also want to offer a very special tribute to the indomitable Mary Lewis. Thank you so much, Mary, for helping us track people's contact details and for all your patient responses to the numerous questions we asked. It is an absolute joy and privilege to work with you!

Besides the contributors, we also want to thank those who provided us with visual memories. The award-winning, internationally renowned photographer Thomas D. Mangelsen very generously gave us access to his entire archive of Jane pictures and Andy Bennett helped clean some up. Mary Paris of the Jane Goodall Institute USA, was very helpful in digging up early Gombe pictures. Patrick McDonnell also gave us a lot of choices for his cartoons of Jane that he drew for his Mutts collection.

It was through Benjamin Beck that our attention was drawn to Salt Water Media. Stephanie Fowler has patiently answered all our questions and assisted us in guiding our little Vessel of Hope and Celebration to the harbor of publication. Thank you, Stephanie!

Marc Bekoff   *Koen Margodt*
Boulder, Colorado   Leuven, Belgium

Help Us Helping Others

Thank you for purchasing this book. All royalties will be offered to projects by contributors to Jane's 90th birthday celebration book. Please check out their work and help us support their efforts for youngsters, chimpanzees, dolphins, rats, bears, cougars, nature and so much more. People, animals and nature need our ongoing support more than ever. Let's celebrate Jane's birthday with acts and donations of hope. Every step matters, every individual matters.

Thank you from the bottom of our hearts!

Marc & Koen

Jane Goodall
Image Source: André Zacher

Some of Jane's travel companions in action
as drawn by Patrick McDonnell for the MUTTS comic strip. (mutts.com)
Patrick is the author of the heartwarming children's book Me ... Jane.

Image Source: Patrick McDonnell

Reflections

Here's what Jane's family and friends had to say about this most remarkable woman.

The Voice in Our Hearts and Heads

— Alicia Kennedy —

In my 2015 contribution to *The Jane Effect*,* I shared how Dr. Jane's messages on the collective difference our individual actions can have resonated with and guided me daily. Here we are, eight years later, and her gentle but determined voice continues to be the voice in my head. Jane's messages permeate my thinking, actions, and emotions. And I know I'm not alone.

Our fragile world is struggling under the challenges of climate change, social injustice, and a broken economic system. But Jane's messages of hope inspire millions globally and catalyse thousands of young people into action through Roots & Shoots. We have become a mighty interconnected network of global changemakers each making a difference through our own "pieces of the jigsaw."

My three daughters, Evelyn, Jane and Megan, grew up in a R&S family where our dinnertime conversations were not about academic or sporting achievements. Instead, we explored the question: "what have we done today to create a better world for animals, people and the environment?" This same question ignited a generation of young changemakers – like Asha Mortel, Asitha Samarawickrama, Jess Pinder, and many others – to do incredible things. Always driven by the lessons of Jane and R&S.

One highlight of my R&S years was joining the inaugural gathering at Windsor. I will never forget that amazing experience of sitting in conversation circles with Jane, Mary, and an assembly of leaders who were transforming their communities across the globe.

---

* My piece from *The Jane Effect* is on my blog page at socialheartedvet.com.

As we celebrate Jane's 90th, I reflect and give thanks for the light she shines for young people everywhere. Jane is the role model for every generation. Thank you, Jane, for inspiring us to combine our skills, gifts and talents with our passions. To follow our own dreams with courage and creativity. And to always, through every action, create a better world for every species and our precious planet. Happy Birthday Dr. Jane!

ALICIA KENNEDY – Veterinarian, Founder Cherished Pets, and Past Director of JGI-Australia.

– Angel, Merlin, and Nick van Lawick –

Cheers to yet another feat! 90! Our generation like to say 'More life' when wishing someone a happy birthday and even though you talk a lot about preparing for your next 'adventure' we wish you more life not just today, but everyday cause of what you bring to our lives.

Being your grandchildren is something that we're still coming to terms with even till today. What a blessing, if we were Buddhists we'd have to have been very virtuous in our past life to be born into such a family. It's a privilege we're proud of and you will always inspire us to do good and care for others.

Grub, Anne Pusey, "Gaj" (Jane), Angel, and Merlin in Gombe, 2023

Image Source: Stephano Lihedule

You have pushed us in ways that you wouldn't even know of. The news of your presence returning to the house when we were younger got us on our toes. You always ask us what we've been up to and our plans in the future not just to know but because you want to help us in any way you can.

Your visits were always a time of reflection and we've always wanted to make you proud in the same way you make us proud every day.

Today we celebrate your life, our beloved GAJ, our symbol of HOPE, our figure of 'indomitable human spirit'.

Lots and lots of love,

Your Grandchildren

A family collage by Nick van Lawick - Grandchildren Angel, Merlin, and Nick are standing behind Jane, holding arms.

Image Source: Nick vanLawick

Happy Birthday, Jane

— Anna Rathman —

Dear Jane,

Let's set the stage ... an evening, formal event at the White House in Washington, DC. An assortment of lawmakers, heads of business, and dignitaries gather amongst the official presidential portraits, gold-leafed wall adornments, and upholstered chairs. In walks the President of the United States, Mr. Joe Biden. As his remarks conclude about honoring contributions of visionary women, it is Women's History month, a long receiving line forms in the adjoining ball room. The crowd begins to disperse and finally it is my time to greet the President. He extends his hand and I, in exchange, extend thanks and greetings from Dr. Jane Goodall.

"Dr. Jane Goodall?! Where is she?" asks the President.

"She is at home in the UK, Mr. President." I respond.

"Well get her on the phone!!" directs the President in the presence of all his staff and remaining guests.

You know the rest of the story. And thanks to Mary Lewis' answering the phone, and quick action, we did indeed get you "on the phone" with President Joe Biden.

While this is a fun, funny, story (especially when we tell it together) what it represents is the power of your global presence. Your wide-ranging positive impacts affect everyone from the highest levels of leadership to the youngest children around the world. That is "Jane Magic".

And as we celebrate your 90th birthday we will all be raising a glass to this Jane Magic and the beautiful ripple effects it makes in our world.

Wishing you a very happy birthday, Jane, and many "magical" more!!

ANNA RATHMANN – Executive Director, Jane Goodall Institute USA.

Jane's Bible

— Anne Pusey —

A joy of my unforgettable years at Gombe was sitting with Jane on the beach of Lake Tanganyika after a day of following the chimpanzees, recounting to her what I had seen. I was studying youngsters and had stories of Atlas nesting away from his mother for the first time; Goblin throwing a tantrum when his mother wouldn't come with him as he followed his idol, Figan; and Little Bee checking out the males next door then returning to visit her mother and sister. Jane was eager to hear about the doings of the chimps she knew so well and unfailingly brought up similar examples from other families that she had observed. Thus, we pieced together a picture of the different pathways to independence.

Fast forward 10 years and I had begun studying the lions of the Serengeti. Jane was steeped in the daily lives of the chimps. She caught me up with news of my young subjects, now adults, when she visited the Serengeti and I visited her in Dar es Salaam. In 1986, her *magnum opus, The Chimpanzees of Gombe* was published. This wonderful book brilliantly combines stories—word pictures—of specific episodes that vividly portray the individuals and their behavior with numerical analysis of general trends. Although she was planning to continue her research, her life was forever changed when she attended a conference to celebrate her book's publication and she heard about the plight of chimps in the rest of the world. As she put it, she went in a scientist and came out an activist.

By this time, I was a tenure-track professor with access to computers, statisticians, graduate students and an enthusiastic body of undergraduate assistants. I offered to help Jane by starting to digitize the systematically col-

Jane with student researchers at Gombe.
Anthony Collins is the third from the left of Jane.
Image Source: The Jane Goodall Institute

lected data so that we could analyze everything from the last 30 (now more than 60) years. Through her ceaseless lecture schedule and her Jane Goodall Institute, Jane has continued to fund and run the Gombe Stream Research Institute and maintain the daily chimp follows. Currently, a consortium of my colleagues conduct research at Gombe and continue the digitization and analysis. Any new student in this enterprise starts by reading Jane's popular books. Then, as they get immersed in the work, *The Chimpanzees of Gombe*, dubbed Jane's Bible, becomes indispensable. The several copies in my lab looked like hedgehogs with colored Post Its sticking out in all directions to mark specific incidents and hypotheses. We would wonder where people had disappeared to until they emerged hours later after looking up something specific and then reading on and on, drawn in by the book's magic.

Much of the research following this book has supported and confirmed Jane's early insights with more data. And the hypotheses she advances in the book continue to inspire new research. One area of particular interest on which Jane could only speculate is which males are successful fathers. Now, with DNA extracted from poop, we are getting the answers. It's thrilling to share these with Jane. And so, through Jane's continued efforts, the research on these world-famous individuals and their descendants, continues.

ANNE PUSEY – James B. Duke Distinguished Professor Emerita of Evolutionary Anthropology, Duke University. Former director of the Jane Goodall Institute Research Center, Duke University.

scholars.duke.edu/person/anne.pusey

Anne at crest of Gombe Rift on Christmas Day, 2014.

Image Source: Anthony Collins

Jane Goodall on the Balcony

— Azzedine Downes —

There are few absolutes in our lives. Still, many people want simple answers to complex questions. Jane and I talk about these questions frequently and I rely on her wisdom to help formulate answers that are always truthful if sometimes challenging to deliver. Why challenging? People who love animals sometimes romanticize life in the wild. Life in the wild can be brutal. The alphas may have a wonderful life but if you are at the bottom of the social structure, life can be stressful or violent. I once shared that observation and was challenged. "That's not true, animals live in harmony" was the response. I told the person that I would have to decide, trust Jane Goodall or rely on your observation; I went with Jane Goodall.

Not every conversation with Jane is about work. Our conversations sometimes begin on a surprising launch pad. We were both in Hawaii for the International Union for Conservation of Nature Congress and planned to get together after one of Jane's presentations. From the lobby of the hotel, I phoned up to say I was here. Jane said to come up. When I got to her room, Susana Name told me that Jane had just gone down to look for me. She also pointed to a note on the table next to the open French doors leading to a small faux balcony. I was confused and said that I had just spoken to Jane but would read the note to see what was going on. I stood next to the open window reading the note. Jane suddenly jumped in through the open window and scared the daylights out of me and I almost pushed her off the balcony! This didn't happen in our younger days; she pulled the prank in 2016!

Most importantly, Jane's consistent message has been never to give up hope. With all the bad news that we are faced with every day it is hard to remain positive. Still, there are actions, no matter how small, that everyone can take to save the planet. Start with hope, plant a flower that bees or butterflies like to visit, and make a pledge to share the planet. Don't search too hard to find inspiration. Listen to Jane and you will feel better about yourself and the planet.

AZZENDINE DOWNES – President and CEO of the International Fund for Animal Welfare (IFAW). Member of the Jane Goodall Legacy Foundation Council for Hope.

ifaw.org/international

Dearest Jane

− Bart Weetjens −

Turning 90, what a blessing, and a what a nice opportunity to thank you for who you are !

I already admired you when I was a teen, breeding all kinds of rodents in my bedroom. On TV there was a documentary featuring you, fearlessly studying chimps in the African forest. It triggered not only my admiration, but also a fascination with the natural world we belong to, and a keen interest in Africa, its overwhelming nature, and inspiring cultures.

Some twenty years later, when I had turned my back to the industry, I was training African giant pouched rats to save human lives from disaster and disease by detecting landmines and TB. Out of the blue, I received a mail from you, Dr. Jane Goodall, what a surprise!

You were planning a documentary series about human-animal relationships, and you wished to feature the HeroRATs ... Yaaayyyy! I was going to meet Dr. Jane in person, what a privilege. And what an unforgettable meeting we had. I drove to the center of Morogoro, to help you and your crew navigate the way to Sokoine University, where my organization APOPO had established its headquarters. You preferred joining me in my car instead of staying in yours. Through our dialogue I immediately felt your trust and our shared values. Realizing that you were a soulmate, I felt goosebumps on my skin, this was clearly beyond words.

Later, you wrote in a letter: "Occasionally one meets a person to whom one feels an attraction on the spiritual level. I am so thrilled we met. So honoured to be a friend. You inspire me." And that inspiration, my dearest Jane, is so mutual. You became more than a friend to me. You also became my

Jane and Mus Mus, an ambassador for the HeroRATs of APOPO. The trained giant
pouched rats help detect and eliminate landmines and tuberculosis.
Jane is an Advisory Board Member of APOPO.

Image Source: APOPO.org

mentor and my moral compass. In moments of doubt about difficult ethical
questions I traveled to Dar Es Salaam to see you and ask for your advice. I
will never forget those tropical evenings on your terrace where we talked and
laughed over a glass of something.

You chose to never lose hope, and inspire so many youngsters through
your talks, books and through the Roots & Shoots program, that inspired
an entire generation around the world to embrace their agency in a compassionate and respectful way. While travelling around the world to inspire

audiences, you carry a HeroRAT along with Mr. H. They really must be good friends by now.

Occasionally, when our travel schedules coincided, we meet each other at an event somewhere. It always fulfills me. Particularly, I remember our encounter a few years back in Brussels. It was early winter, we stood at the entry of the house where you were staying, and there was a mistletoe hanging there. You hugged and kissed me goodbye, and as we stood there embracing you said with your trademark quirky smile and complicit gaze: "Look where we are standing, Bart. If only I were forty years younger..." I was, and still am, deeply moved by this. And I can tell you now, I would have said: "Yes Jane, of course!"

BART WEETJENS – Zen Buddhist monk and former product designer who founded APOPO, an organization that trains animals to sniff out landmines and illnesses like Tuberculosis.

bartweetjens.be

Celebrating the Remarkable Journey of Dr. Jane Goodall on Her 90th Birthday

— Bawa Jain —

A Very Happy Birthday my Dearest "Jane" Sister! As the world comes together to celebrate the 90th birthday of Dame Dr. Jane Goodall, it is not just a commemoration of years lived, but a tribute to a life dedicated to the preservation of our planet and its diverse inhabitants. To me, she is not only a global icon and a pioneering scientist, but more importantly, I am honored to have her as a sister, a living saint whose profound impact has resonated with me for over two and half decades. From the moment I first encountered her, her wisdom, dedication, and compassion have left an indelible mark deeply etched on my soul.

Over the span of 20 to nearly 25 years, I have had the privilege of looking up to Jane Goodall as a guiding light, an inspiration that has shaped my perspective on life and our role as stewards of the Earth. The honor of calling her my dear sister is not just a title, but a reflection of the deep bond that forms when one is touched by the selflessness and grace that radiate from her being.

Dr. Jane Goodall's legacy is woven into the very fabric of conservation history. Beyond her groundbreaking research on chimpanzees, she has redefined the contours of species conservation, recognizing that the intricate web of life extends beyond individual species to encompass the well-being of local communities and the delicate balance of ecosystems. Her holistic approach has illuminated the interconnectedness of all living things, emphasizing that protecting our environment is inseparable from ensuring the welfare of the people who inhabit it.

The day we first met, in the vibrant city of San Francisco during The

State of The World Forum, remains etched in my memory. While dignitaries and world leaders convened within the walls, Jane Goodall and I found ourselves seated on the steps outside the Fairmont Hotel and Grace Cathedral, immersed for hours and engaged in a dialogue that transcended the grandeur of the event. As we discussed the challenges that confront our world, her words carried the weight of experience and unwavering determination with a youthful vigor.

Dr. Jane Goodall's impact extends far beyond her words. She has ignited a global movement, inspiring individuals of all walks of life to take action and make a positive change. Her advocacy for environmental conservation, education, and the power of empathy has sown the seeds of transformation in countless hearts and minds. Her passion for Roots and Shoots is now impacting the world and empowering youth across the globe.

As we celebrate her 90th birthday, let us reflect on the profound legacy of Dr. Jane Goodall. Her life's work serves as a beacon of hope in an increasingly complex world, reminding us of the urgent need to protect and preserve our planet for future generations. Her journey exemplifies the profound impact that one person can have when driven by passion, purpose, and an unyielding commitment to a better world.

In honoring Jane Goodall on this milestone occasion, I join countless others in expressing deep gratitude for her contributions. She is not just a sister and an inspiration; she is a living Saint, a testament to the potential of humanity to create positive change. As we celebrate her 90 years of life, may we be inspired to continue her legacy of compassion, conservation, and hope.

Wish you the very best for the next decade my dear Sister. The words that come to my mind aptly express my sentiments of where I see you currently!

The woods are lovely, dark and deep,
But I have promises to keep,
And miles to go before I sleep,
And miles to go before I sleep.

– Robert Frost

BAWA JAIN – Secretary General, World Council of Religious Leaders. Founder: Centre for Responsible Leadership

– Benjamin Beck –

Happy Birthday Jane. Let's recall your first scientific talk, in 1962 at a symposium organized by the Zoological Society of London and chaired by the eminent primatologist Sir Solly Zuckerman. On that morning, April 12th, you presented a paper titled "Feeding Behaviour in Wild Chimpanzees: A Preliminary Report," a summary of your master's thesis. With what we have come to know as your exacting understatement, you stunned primatologists with descriptions of several forms of tool use and tool making by the wild chimpanzees of the then-Gombe Stream Reserve in Tanzania, and identified their tool behavior as "a social tradition, which represents ..... a primitive culture." You also documented the chimpanzees eating, reluctantly sharing, small mammals. Media accounts of the session sent philosophers, and professors and their students (including me, a college senior at the time) searching for new concepts of tools and humanity. All of your observations have been verified.

Zuckerman assigned himself as the session's discussant, even though it can fairly be said that Zuckerman got more wrong than right in his career as a primatologist. His concluding remarks on this day were patronizing and arrogant, a masterwork in the art of snark. He used the pronoun "I" 55 times in five pages of what were intended to be a review of others' contributions. He characterized the work as "anecdote and speculation, much of which departs to no little extent from the source out of which it is generated," making it clear he was referring to one particular paper. Could he have been implying that Jane had exaggerated or fabricated her findings? "Miss Goodall's [might have been] a special group of chimpanzees living under very favourable conditions, plagued by few predators and enjoying abun-

Jane climbing a tree to get a better view of chimpanzees.

Image Source: Hugo van Lawick

dant food." Later he concluded "[I] hardly think that any of the examples that have been given about occasional meat-eating in monkeys and apes alter the basic fact that that these creatures are predominantly vegetarian and frugivorous." Although Goodall made no mention of dominance and sex ratio of the Gombe chimpanzees (at least in the print version) Zuckerman criticizes her for disregarding "a considerably wider range of material and observation" that would predict that there would be more adult females than males in a wild primate group, and that adult males would not be sharing food with other group members.

The world is grateful Jane that you were undeterred by Zuckerman's conceit. We celebrate your tenacity and kindness.

BENJAMIN BECK – Primatologist and expert on tool behavior

Cited Articles

Goodall, J. 1963 Feeding behaviour in wild chimpanzees: A preliminary report. The Primates; Symposia of the Zoological Society of London, Number 10, pp 39-47.

Zuckerman, S. 1963 Concluding remarks. Ibid, pp 119-123.

Just Jane

— Brad McLain —

A few years back, I was a new board member for the Jane Goodall Institute. But I felt lost and unsure of my place. I often chose to hold back rather than rock the boat with ideas that were contrary to more experienced members. I was considering stepping aside. But one night I had a transformative experience with Jane that changed everything.

While Jane was in town as part of a tour of the U.S., she called me to announce she was coming over for dinner (phrased much more graciously of course). Knowing she is the frequent guest of the rich and famous, I immediately became self-conscious about my private little world. I explained to her that our house is quite humble —the home of two young children, overly-affectionate dogs, and an unremitting mess that migrates around. "Perfect!" she said. "We'll be there in thirty minutes."

I panicked into action. Get food, drinks, clean the house, feed the dogs, shower, and oh my god, we have leather couches and the world's preeminent animal rights activist is coming to visit! It's amazing what runs through your mind in such moments.

After dinner, we all settled on the back porch under blankets and a distant October thunderstorm. We swapped stories and pondered the ultimate meaning of our lives for hours. I mentioned how I was feeling self-conscious about my smallish and untidy home (and life). "Who do you think I am, Bradley?" she said to me. "I raised my son in a tent! In the jungle!" Ever the kind heart, she added, "This is my kind of house."

She explained that she didn't know who this 'celebrity Jane Goodall' was. She was deeply touched by the public's reception, but the fame and legend were "not my fault," she said. "I discovered that I can speak effectively on behalf of the animals I love so dearly, because they cannot speak for themselves. But this famous Jane Goodall 'creature' is not the real me. I'm Just Jane."

Later as we were cleaning up, we noticed that Jane had gone inside earlier and never returned. At last we heard my young son Kai upstairs saying, "Come in a little farther to see the spiders!" Jane was on hands and knees with a flashlight, crawling into a secret fort under his bed.

"I don't mind the spiders," she said. "I just don't think I can turn round again if I go all the way in!" But she did – of course she did.

It was a surreal and magical night.

Whatever I had previously thought about the legendary Dr. Goodall from growing up learning about her discoveries in Gombe, to the dozens of books and documentaries about her, to her work as a UN peace messenger, and her mythologized pope-like status as a symbol of hope – beyond all that, on this night, I met "Just Jane." And doing so unlocked "Just Brad."

I went from passive board member to contributing everything I had no matter how humble. I served on the executive team for the Jane Goodall Institute, led fund-raising events, chaired the Roots and Shoots leadership committee, and still today include Jane's important work as part of everything I do, from speaking engagements to neighborhood pizza parties to books like this. I'm currently in production for a new documentary called "Just Jane." This is what Jane does. She inspires. And that wonderful night with "Just Jane" changed me.

BRAD MCLAIN – Author of *Designing Transformative Experiences*. Social Scientist @ CU Boulder and NCWIT. Longtime JGI collaborator, past board member, Roots & Shoots Evangelist (Excerpted and adapted from Dr. McLain's book, *Designing Transformative Experiences*, BK Publishing 2023.)

www.DesigningTransformativeExperiences.com

bkconnection.com

A Salute on Your 90ᵗʰ Journey Around the Sun

– Sir Brian May –

J ane Goodall is beyond doubt the greatest reformer of Scientific ethics of our time. It's hard to define the point in History when Science became divorced from Art, and from Morality. But at the time when Jane was working on her thesis, it was the norm in scientific research not only to ignore the dignity and feelings of non-human animals, but to brand any suggestion that animals were sentient beings as unscientific – a weakness which would undermine any scientific enquiry. Jane came directly up against this kind of thinking in her wish to give the animals she was studying names instead of numbers. She was directed to change her mind, or have her research discredited. As is now well-known, in resisting this directive, and ultimately overturning this horrible misconception, she struck at the very heart of the blind cruelty which had characterised Science for at least 150 years. By her example she influenced, and continues to influence, decisions in Scientific enquiry all around the world. Her subsequent life, dedicated to continuing the process of the emancipation of animals, has inspired all of us who believe that the old, permitted inhumanities must be expunged from any decent society.

I'm proud to be one of Jane's dedicated supporters, and to help carry her message into the future. I have learned that *Anthropomorphism* is a deeply suspect word, used to defend cruelty to creatures unable to speak and defend themselves against Human exploitation. And that *Anthropocentrism* is at the root of all abuse of our fellow creatures on Earth – the logically unsupportable belief that Humans are the only species on the planet worthy of consideration. Every day, every one of us who toils along the long path towards making the Human Race behave decently, is inspired by Jane's vision, talent, and tenacity.

Dear Jane, we all salute you on completing your 90th journey around the Sun, and wish you many more years of joy and fulfilment.

SIR BRIAN MAY – CBE, founding member of Queen and founder of the Save Me Trust with Anne Brummer.

savemetrust.co.uk

What The Animals Say
When Jane Comes 'Round

— Cara Blessley Lowe —

" Shhhh" they say. "She's here!"

"How can you tell it's her?" the youngest asks.

"The ponytail, always the ponytail" the teenager says. He's cheeky. (Some call him the 'know-it-all'.)

"Does she ever let her hair down?" another asks. She is quiet, and still has the faint markings of a fawn though her long legs are strong now, and she no longer wobbles.

A husky voice replies from under the cover of a baobab limb. He is wizened, grizzled, with a grey chin and eyes with deep creases at the corners from many hours spent watching. "I have never seen her without her ponytail. Ever."

The forest falls to a hush. Impossibly, not even a branch cracks or a fir needle whispers beneath Jane's steps. She moves through the dense underbrush like an angel, weightless. Even the air is still.

A bee eater flies in for a closer look and for the first time, the animals notice both the tree and the bird. (They are, both of them, beautiful in their own way.)

Jane doesn't do much. She doesn't have to – she already has. She gathers her legs beneath her body and settles into the earth. Jane is there to be with them, and that is enough.

CARA BLESSLEY LOWE – Writer and script supervisor. She and Tom Mangelsen founded the nonprofit Cougar Fund in 2001 following the nationwide broadcast of her documentary short, *Spirit of the Rockies.*

Celebrating 25 Years of Friendship and Conservation Partnership

— Charles Knowles —

What a joyous occasion it is to celebrate Jane and her 90th birthday! She has made such a profound difference for the world and she's touched so many lives, including my own.

I had the good fortune to meet Jane nearly 25 years ago when I was setting out to start the Wildlife Conservation Network. Not only was she instrumental in helping me connect with my co-founders, she graciously offered to speak at our annual Expos which resulted in us doubling our attendance every year she came. Her partnership has helped WCN grown to the point that we have now deployed over $300 million to support wildlife conservation around the world.

Last year we celebrated the 20th anniversary of WCN and Jane graciously joined us again. She greeted me on stage with a bottle of Scotch and we proceeded to have a nip together in front of 1,500 audience members. Jane also presented me with a candle. She asked for a light and several lighters and books of matches were thrown up on stage. As I proceeded to try and light the candle I realized that it was an electric candle, a fact that brought much laughter from the audience. It was a wonderful and inspiring evening and yet another example of the magic that Jane brings wherever she goes.

Jane has been a great partner in conservation, but also a great friend. I'm now a father of 11-year-old twin boys and they have grown up knowing Daddy's friend Jane. In fact, when they were in first grade, they asked their teacher if they could invite a friend to come and speak to the class about

Jane with Charles Knowles and family at their sons' school.
Image Source: Charles Knowles

animals. You can imagine the teacher's reaction when she learned that it was Jane Goodall who would be coming to class!

Thank you, Jane. You touched my life when I was a little boy and you continue to have a huge impact on me today. You inspired me and helped me start WCN and you continue to inspire people to get involved in saving wild animals and wild places around the world.

Jane, it has been, and continues to be a wonderful journey together. Thank you so much for your vision, passion, leadership, but most of all, your friendship.

Much Love,
Charlie

CHARLES KNOWLES – President and Co-founder, Wildlife Conservation Network.

wildnet.org

Jane on Capitol Hill

– Chris Heyde –

I don't think I have ever been in Dr. Goodall's (I know she will read this and shake her head because I have never been able to call her Jane despite constant requests) presence without some admirer saying, "you have always been such an inspiration to me" and that couldn't be truer for me as well. I am lucky to have taken that inspiration even further to have had the privilege of working with her for about 25 years on a variety of legislative issues before the US Congress and government. Like so many my age who only had a few channels to watch, I grew up watching Dr. Goodall's National Geographic specials and was captivated not because she was a woman doing what she did, but that she was an expert bringing something precious and unknown to families on a regular basis in an understandable way.

Fast forward to the late 90s, as a Senate staffer, I was invited to attend an event on Capitol Hill featuring Dr. Goodall talking about her work with chimpanzees to legislators. Not often you can pinpoint the exact chance event that kicked off a life changing career change, but this encounter was the moment for me. Not only was Dr. Goodall present, but I also got to meet her friend, Christine Stevens, who is considered the mother of the animal welfare movement. Like Dr. Goodall, Mrs. Stevens always wore a ponytail in her gray hair so how could go wrong in admiring their passion. Both are giants in the field of animal protection and conservation, and I now had contact with individuals who to this date guide my beliefs. I should quickly add that I went on to work for Mrs. Stevens and her group for 18 years.

Flash forward two years and I was back in contact with Dr. Goodall asking for her support in passage of a bill called the CHIMP Act, which created a national chimpanzee sanctuary to retire chimps no longer used in biomedical research. She never hesitated a minute to join in that effort and our work culminated in the passage of the CHIMP Act which has allowed for many (unfortunately not all yet) chimpanzees to be free of cages in research facilities. She agreed to come and testify on behalf of the CHIMP Act and despite suffering a flare-up of malaria, once that door to the hearing room opened, she was treated and respected as the world-famous expert that she is. Rarely is legislation passed so swiftly, but thanks to testimony from the leading expert on chimpanzees, the CHIMP Act was resoundingly approved and continues today.

There isn't enough space to talk about all the support and time Dr. Goodall has given to me on various legislative initiatives over the years, but I am always happy to brag that I know Dr. Goodall and she is as amazing in person and life as she was to a kid watching the various tv specials that shaped generations views of the environment, wildlife and especially the amazing chimpanzee.

Happy Birthday Jane!

CHRIS HEYDE – Founder/CEO Blue Marble Strategy

How a Job Interview 22 Years Ago
Changed My Life

— Claire Quarendon —

I have had the pleasure of working with Jane for over 22 years. She is a most remarkable woman and my respect and admiration for her has only grown over the years. Jane travels over 300 days a year, visiting countries all over the world, sharing her 'Reasons for Hope'. How she has the energy to

Flying Giant Peace Doves, 2005
Image Source: Thomas D. Mangelsen

keep this pace up well into her eighties is beyond me. When you suggest she slows down, she says, 'I can't I have more to do and less time to do it in.'

When I lost my father, 16 years ago, Jane was a wonderful support. She had met dad many times over the years I have worked for the Jane Goodall Institute UK and he also admired what she has achieved over the years. Dad was more than happy to help make and "fly" Giant Peace doves, as were various other family members, on Roots & Shoots Day of Peace which is held every year in September.

Although I have heard a lot of Jane's lectures over the years, each one is fresh and slightly different in what it focuses on. When Jane talks, everyone else in the room disappears and you are taken on an amazing ride. Jane never uses notes, which is remarkable for someone who talks for just under an hour. I have never seen/heard Jane stumped when asked the many questions that are put to her at events and on the radio and tv.

I have the pleasure of seeing Jane several times a year and there is always a warm welcome for me at 'The Birches', her family home in Bournemouth, from Jane and her sister Judy.

CLAIRE QUARENDON – Office Administrator, Jane Goodall Institute - UK.

Jane Goodall and Her Love for Lobsters

— Craig Foster —

As a young child I grew up seeing rock lobster every day. Huge numbers lived in the great kelp forest next to my house. I love how these animals teach by showing me curious parts of their lives, forcing me to go on long journeys to understand how their intricate life cycles unfold. Going into their inner worlds and then seeing them being caught in traps by the thousands sends a deep chill through me. If one in a million survive to adulthood and we take so many adults, what chance have they got as a species?

Coincidentally I'm reminded of a lobster poem that my friend Jane Goodall recently sent me, which she wrote when she was only 13. What insight she had even when so young, what empathy to see the world from a crustacean's perspective. I've seen with my own eyes, just in my lifetime, the collapse of our rock lobster population. Overfishing has led to it being decimated to less than 2 percent of its original number. How is it that a child knows things that are somehow not privy to our strange adult minds? Is it because the child's mind is still wild, and the tame adult is lost to economic coercion?

Lobster

A lobster, caught inside a trap,
Simply could not find a gap
And gazing sadly at the prawn
Remarked 'Why was I ever born?'

The prawn unable to reply
Just heaved a sympathetic sigh
And only wished he knew
Something useful he could do.

'No matter how I snap my jaws
'Or tear the meshing with my claws
'I simply cannot find the gap
'Through which I entered to this trap.

'And so I'll simply have to wait,
'Trembling, to meet my fate
'My life cut short by many years
'And nobody to heed my tears.

'Most humans think we don't feel pain
'And those who know speak out in vain
'And so I'll suffer - boiled to death
'For humans to enjoy my flesh'.

The prawn looked mournful as prawns will
And slowly waved a feathered gill
But could not think of any plan
To save his cousin from the pan

At last the trap commenced to rise
The pawn began to wipe his eyes
And having cried a little cry
He sadly waved his friend goodbye.

The mesh which kept the lobster in
The prawn slipped through he was so thin
And tearfully he waved a claw
Before he swam towards the shore

The lobster travelled to his fate
And soon appeared upon a plate
His erstwhile black had changed to red
His body lifeless and quite dead

Post Script

The fat man sat opposite his lover.
They bit into the white flesh
And swallowed the fear and pain and death
Into their own bodies. And smacked their lips
And paid the bill and went to the opera.

Even as a young child Jane was already working as an animal advocate, providing a voice for the voiceless ones.

CRAIG FOSTER – Co-founder Seachange project and known for the film *My Octopus Teacher*.

seachangeproject.com

Animal Advocacy in the Anthropocene

— Dale Jamieson —

Jane Goodall is larger than life: scientist, conservationist, peacemaker, mentor, and role model to millions of people all over the world. Born in London in 1934, she was not keen on going to school, instead dreaming of nature, animals, and wild places. She explored the garden, wrote poetry, read philosophy, and felt close to God. Greatly influenced by *Dr. Doolittle* and *The Jungle Book*, Edgar Rice Burroughs swept her off her feet with his Tarzan stories. When she arrived in Kenya in 1957, the Mau-Mau rebellion had only recently ended and Nairobi had little more than 200,000 people. Gombe was a small colonial hunting reserve in a vast ocean of trees. Today Nairobi has upwards of 4 million people, and Gombe Stream National Park is a small green island surrounded by farms, fields, and villages. Since Jane began dreaming of Africa, its human population has increased by a factor of 6 while the wild chimpanzee population has decreased by a factor of 5. The causes include growing affluence in much of the world, the globalization of consumption and supply chains, and the radical inequality that creates destructive and even insane incentives. Climate change is also beginning to be felt, and new patterns of infectious disease are taking shape. The romantic world of Jane's childhood has disappeared and is becoming further out of reach all the time.

Although she was unconventional in many ways, Jane made her name as a scientist in her first quarter century in Gombe. Her 1986 book, *The Chimpanzees of Gombe: Patterns of Behavior*, effectively founded a field. It has been cited more than 8,000 times.

But after the publication of the book: Jane came out as an advocate:

first for chimpanzees, then for conservation, then for all kinds of suffering animals and humans. The precipitating event was the 1986 "Understanding Chimpanzees" conference organized by the Chicago Academy of Sciences. Jane was shocked to discover how rapidly chimpanzees were vanishing across Africa, and angered by how they were treated in laboratories in North America and Europe. Because she was now a global celebrity and the world's leading chimpanzee expert, she had access to leaders and institutions normally shielded from examination and criticism. Her decades of advocacy have made huge differences for chimpanzees far beyond Gombe, and for aquatic animals, farmed animals, captive animals, and many of the world's people.

Jane's most remarkable quality (putting aside her indomitable optimism) is the way she embraces change (who becomes vegan in their 80s?). Jane is not the only scientist of her generation who went from studying animals to advocating for them (George Schaller and Paul Ehrlich come to mind), but no one has embraced such a wide range of causes with such passion and commitment. As someone with a mild case of prosopagnosia, I can't even imagine how difficult it must be to spend 300 days on the road, usually surrounded by adoring strangers and unrecognizable friends.

Jane's love of animals and nature was born in the Holocene but she has become their most important advocate in the early days of the Anthropocene. Since Jane was a little girl dreaming of Africa, human/animal relationships have inverted. Today less than 2% of the biomass of terrestrial mammals is composed of free-ranging non-human mammals, and even for the wild few their behavior and opportunities are severely restrained and restricted by human action. Rather than a vast wilderness filled with mysterious threats, nature itself is under threat from humanity, even as we threaten ourselves. We are clumsy giants straddling the globe, destroying things we barely notice, even when we have the best of intentions. Because of our power and numbers, the personal and intimate virtues of the Holocene— love, respect, and empathy —are not adequate to the present moment. They will have to assume a different shape and form in the Anthropocene. Jane has guided us from one epoch to another, developing forms of advocacy that have remarkable power in different times and places. But the greatest honor we can give to Jane on her 90th birthday, inspired, motivated, and energized by her example, is to make the work of developing new strategies and ways of life our own. The imperative to protect our home and all of its inhabitants remains.

DALE JAMIESON – Director, NYU Center for Environmental and Animal Protection. Affiliated Professor of Law, Medical Ethics, and Bioethics. Professor of Environmental Studies and Philosophy Emeritus.

Jane's Friendship

— Dan Miller —

Before I knew Jane, I thought she was some crazy woman working in the jungle with Chimps. Well yes, she is, crazy, as I learned in our 20-year friendship, but in the best possible way: she is crazy about wildlife, people and the environment.

I was first introduced to Jane through her amazing Roots & Shoots program. JGI received a grant to establish a R&S program in South Dakota, with emphasize on rural communities and Native Americans. Jane inspires me to keep this programming going, and after 20 years, it continues to adapt to our changing audiences. I live daily by a phrase I learned from Jane: "Every individual matters and everyone can make a difference."

I have been blessed to travel with Jane across the prairies of South Dakota. She truly loves the landscapes, wildlife, and the people. I remember driving with her in the Badlands on a scorching summer day. Imagine our relief when we were finally able to find a small shaded area—hard to come by in the Badlands –in which we could enjoy a picnic together. We also drove to Pine Ridge Reservation and met with the elders. They were immediately amazed by her spirit and kindness, special traits of her that inspires both me and millions of her fans. Many of these unusual events seem to happen when Jane is around.

Jane is one of the greatest storytellers I know. Listening to her lectures has always inspired me to improve myself both personally and professionally. After her lectures, she always signed books, and people had their photos taken with her. Many of them went away crying, just as I did, happy to have a memento of their time with her. I still enjoy the many photos I have taken of Jane.

When I was very sick with West Nile, Jane kept in touch with me often. Once again, her inspiration empowered me, this time to get better and keep on going. This is one of my most memorable experiences with her.

Jane's kind soft words continue to resonate with me in my daily work with animals and people. If only we could find more Jane qualities in the world, during these tough times.

Happy 90th birthday.

Love and peace.

DAN MILLER – Director of Bramble Park Zoo

In the Footsteps of Jane Goodall

– Daniel Hänni –

Already as a child I was interested in apes and monkeys. I spent a lot of time in the forest in trees. Most of the time I went to the forest alone, dreaming of living in the jungle like Tarzan and being one with the animals and the forest. I climbed the beech trees because these deciduous trees reminded me most of the jungle giants. I often climbed up to 15-20 meters, enjoyed the cool breeze and was dreaming of the rainforest.

Around 1977 I came across the articles about Jane Goodall and the chimpanzees. At that time I already decided to do something like that. In 1980, as a 12-year-old, I wrote in a school essay that I would live in the jungle and stop the deforestation of the rainforests. I would find a lost treasure and use the money to reforest the jungle. But many years passed before that happened. First I focused on my Track & Field career and was in the real estate business. When I decided to go back to school and to university to study anthropology, I finally felt ready to support Jane. In 2004 I founded the Jane Goodall Institute Switzerland's office and met Jane for the first time at the Global Meeting 2005 in Florence. For me it was like meeting a Selenium relative. I think we had similar thoughts and longings in childhood, maybe even similar desires that one has in young childhood.

I felt understood by her without many words. In her own special way, she gave me the courage to do something I didn't expect to be able to do. If I had known at the beginning what I would be facing, I probably would not have had the courage to open the office in Switzerland. But you grow with the challenges, and I knew inside that Jane would give me that strength. Even today I feel her energy inside me, driving me to keep going.

Today, over 40 years after I wrote my school essay, we are planting over 1 million trees each year in Uganda in the corridor project between Budongo Forest and Bugoma Forest for the chimpanzees. Although I didn't know it at the time, you, Jane, made this possible for me. You blazed the trail for me that I had wanted to follow since I was a child. So, in my own way, I was able to follow in your footsteps and achieve something that was already dormant in the little boy at that time. Thank you Jane, for making my life a special one too.

Love,
Dani

DANIEL HÄNNI – Executive Director and founder of the Swiss chapter of JGI

janegoodall.ch

The Great Hug

– Daniela De Donno –

A day at JGI Italy.
"A communication from Jane, a new campaign' I tell Lara, 'Don't worry... we will make it."

"For?"

"*Dolphins.*"

"Essential. I love dolphins!"

"Me too."

"Most people do, I guess."

"Let's start with socials: JGI calls for a ban on the capture, keeping and breeding of dolphins for captivity."

"Remember they wanted to host Jane for the opening of an aquarium with dolphins a few years ago around here?"

"How awful. We didn't even ask Jane, who would ever want to see a dolphin locked up in a tank."

"I would say spread the campaign on dolphins after the petition against the use of cages in *intensive farming*. Jane is the godmother, and we collected signatures."

"Well, first the cages. Unfortunately, we have those pictures of the *chickens* and *pigs* that are so cruel, but everyone has to understand. After a few days we will deal with the dolphins."

"And the news about the *Sanganigwa children* and the *dogs*?"

"Meaning?"

"Do you remember the problem in Kigoma, at the *orphanage*: the children showing little familiarity with the dogs kept in the compound for security..., mostly the children were just afraid of the dogs and showed no interest or curiosity in them."

"...and we asked Jane for advice."

"Correct. Jane had suggested that we raise a puppy together with the children, that they be together from the start. Following her advice, we got the puppy and look at these pictures, what beautiful pictures! Even too much love now!"

"Jane will love the pictures of the kids so happy scrambling puppies."

"Where are we with the *wolf* festival?"

"Jane won't be able to be there but she definitely wants to send a video message for the wolf."

"The one about the *bears* in the Abruzzo Park was also significant."

"Yes. A R&S group wrote, they want to learn the songs of our sung fable about the *northern bald ibis*. It is a school along the migration route where Jane came to meet the children to talk about the ibis."

"Great!"

"They're organising a choir and a video to broadcast."

"Oh, and then R&S Beijing and R&S Malaysia asked to support them in fighting the crime of *illegal tiger trafficking*."

" Yes, I read Jane's statement. It needs to be translated and then spread on socials. There is also a quiz: how well do we really know the tiger?"

"Before leaving, we have to send Antonio the subtitles for his video on Jane and you for the 25th anniversary of JGI Italy."

"I'll do that. Do you remember where we were at?"

"When talking about R&S's *mobile phones recycling campaign* in Italian schools and the devastating impact in DR Congo caused by the mining of minerals to produce mobile phones. We got to that screenshot in the video where Jane says: not just for *chimpanzees*, for *gorillas* too. This needs to be explained.'

"Finished."

"Bye."

"Nooo! The *salmons*, we were forgetting the salmons, the salmons from the intensive farming!!!"

DANIELA DE DONNO – Founder and Executive Director of JGI Italy

A Trip to DR CONGO: Hope in War Time

– Dario Merlo Kasuku –

In November 2012, war raged in eastern DRC. The rebels are at the gates of the city of Goma where the headquarters of the Jane Goodall Institute in the Democratic Republic of Congo was located. I was the program director and a trip for Jane was planned during this same period. During the discussions to prepare the trip, everyone strongly advises against a visit in this context, the British Embassy prohibits any non-essential visit to this part of the DRC but Jane insists on coming. When asked for my opinion, I describe as faithfully as possible the major risks, the chaos we are experiencing and I express my concerns but to my surprise, Jane insists! She wants to come "to understand the context in which we work."

When you are part of the big family of the Jane Goodall Institute, you know Jane's strength of character, her love for people, her desire to change things for the better, but no one really measures her courage. She was willing to risk her life for us.

Jane came to the DRC in the middle of the war. During the speech in one of the largest universities in the region, in front of a crowd of more than 3,000 students gathered for the occasion, the sound of bombs and heavy weapons resounded. That day, Jane spoke with determination about peace and our ability to make this world a better place by being the agents of the change we want to see. The war was raging less than 15 km from the city. The conference lasted 2 hours and we had to cut it short, put Jane in a sheltered hotel by the lake. From there, it could still have been evacuated if the fighting were to take place in the town of Goma. I proposed (rather – obliged) to Jane to leave that day and to cancel the big Roots & Shoots gathering which

was planned for the following day but she replied that it was out of the question. She would leave if I decided to leave too. I took her in my arms and told her that I loved her.

The next day, during her speech in front of thousands of Roots & Shoots members, Jane was there with the inseparable Mr H (her little stuffed monkey that accompanies her everywhere). The fighting had come closer, a shell had fallen in the morning near the airport of the city, located 2 km from us. Jane spoke to young people, reassured them about their future by telling them about her childhood, her journey and how she overcame obstacles. She talked to them about following their dreams and continuing to believe in them.

That day, Jane awakened in me the desire to follow my dream. That of building an international school in eastern DRC that would train a generation capable of rebuilding this country. I realized this dream in 2017 with Kivu International School (kisdrc.net) and since then I have been witnessing the miracle of Education.

Jane returned to another destination after meeting and inspiring thousands of young people, meetings with the Minister of the Environment and other political actors, visits to a park and a center that recovers illegally detained chimpanzees. The rebels took the town a few days after she left.

I understood that day that she is animated by an invisible force that transforms the world.

DARIO MERLO KASUKU – CoFondateur & Président du Conseil d'Administration de Kivu International School. Fondateur de Promo Jeune Basket.

kisdrc.net

promojeune-basket.com

Hope is Her Response

— Dave Matthews —

I first met Jane Goodall on July 7th, 2007, at Giant Stadium. We were both scheduled to appear at the Live Earth concert which was to be broadcast from cities around the globe.

Moments before Jane was due to go on stage, I introduced myself. As a boy I'd followed her through the pages of National Geographic and on television as she braved the jungles of Tanzania pursuing a greater understanding of wild chimpanzees. Here I was meeting her in person. I was awestruck but she greeted my fawning with quiet grace and her familiar smile.

Jane walked on stage to little fanfare. I watched from the wings as she stepped up to the mic alone and confessed to the audience of nearly one hundred thousand that she wasn't accustomed to addressing so many. So, she said she would begin with the call a chimpanzee would make to greet their fellow chimps across the jungle. She followed with her now famously convincing impersonation of a chimpanzee greeting.

The crowd was at first silent but soon Jane's strange, haunting call seemed to elicit some primal response that rose to a deafening, joyful roar. This slight but mighty woman from Bournemouth in England blew the roof off a stadium with nothing but her voice.

I get goosebumps now as I remember that moment. I'd known of Jane from when I was a kid. My parents were passionate about wilderness and conservation and people like Jane Goodall were heroes in our house.

The images of this young woman in the jungle reaching across millions of years of evolution to communicate with the wild chimpanzees were hard to believe. It was the stuff of our wildest dreams. We have her to thank for how profoundly our understanding of primates has evolved.

Before Jane's work there was a clear line among scientists between humans and the rest of the beasts of the earth. The footage of chimps using tools of their own that Jane brought back to England blurred that line. It was with this evidence from Gombe that the title "man the tool maker" was laid to rest.

We now know that other animals, from crows to rats to elephants, use tools taken from their surroundings. Still the similarity between chimp and human behavior is particularly profound.

Since the day we met at that concert in 2007, I have been honored to share the stage with Jane many times, but even more than that I've enjoyed her company. If you plan on taking a walk with Jane, be prepared to go off the beaten path, crawl through a few thorny bushes, sit on the ground to rest, laugh, and be amazed.

Whether it's watching a raccoon at the bird feeder, or half a million sandhill cranes filling the Nebraska sky with wings and cries or crouching next to a chimp in the jungles of Tanzania, Jane's childlike wonder at the world's wilderness and its occupants is infectious and inspiring. Her commonsense commitment to its preservation is absolute and urgent. It is a commitment to the whole earth.

Our species' endless hunger for more power and more money has only one outcome. The planet is so abundant, but it is finite, and we are quickly reaching its end. Our oceans and forests cannot sustain us at this rate. It seems we haven't yet learned, despite Jane's efforts, that we rely on a healthy earth for our survival.

Jane remains hopeful. Her hope does not deny the direness of our situation. But hope is her response.

Jane builds a good fire with ease and enjoys a little red wine. She is an absolute vegan and doesn't tolerate rinsing dishes before putting them in the dishwasher. She doesn't understand high heel shoes and likes Johnny Walker. Jane is a world leader in conservation as she has been for my entire life. Jane can speak to animals, and she is my friend.

Let's hope her hope is well founded.

DAVE MATTHEWS – Singer and songwriter

The Jane Goodall I Know: My Memories and Connections to Dr. Jane Goodall

— Emmanuel Mtiti —

During my early childhood around the 60s, I grew up in Kalalangabo village, a third village south of Gombe National Park. Here is where as a child I first heard stories about the strange animals that live within the Gombe forest. I then realized that some young people from the village who had been to school got employed as research assistants for the chimpanzee research in Gombe. They came back with stories on using cameras to track chimpanzees and gave real stories on chimpanzees' behavior and accounts of their life in the forest. They were admired by the members of our communities for their advancement in wildlife knowledge. We often hear about a "white lady" (mzungu) who lives in the Gombe forest (mostly known to us as Kasekela) following the chimpanzees and has employed a number of people to help with the research. The whole story appeared bizarre to local communities since there was not much we thought could be learned from the chimpanzees and other wildlife. The wildlife was mainly the source of bushmeat since the region did not have much livestock due to high tsetse fly infestation.

During those days we used to see chimpanzees and hear their calls even in forests closer to the village and beyond, but they were not the subject of traditional hunting but could be accidentally killed by snares intended for other wildlife.

Despite growing up closer to chimpanzees I never thought of becoming a researcher or a conservationist since my ambition was in health since poor health services were one of the major problems and became a priority to rural communities. My family later moved from the village where my father was engaged in fishing, and he once had a fishing camp within Gombe beaches. The further we moved the more we lost touch to stories of Gombe and the researchers.

After completing my studies, I was employed by the Ministry of Health where I went through a number of positions the last one being coordinating preventive services in the region. It was during this time again that I really came closer to Jane and George Strunden as they were introducing the new environmental conservation project. George was the first Project Manager for the Lake Tanganyika Catchment Reforestation and Education (TACARE) project.

Jane was on her regular visits to Kigoma and he used to visit the regional hospital where she often had talks with the Regional Medical Officer (the late Dr. Mbaruku). Whenever we met, we discussed issues of common interest specifically the conservation activities and health of the people around

Jane and Emmanuel overlooking the TACARE project.

Image Source: Richard Koburg

what was by then the Gombe National Park. This is the time we discussed issues of common interest and I started realizing (after learning from Jane) how we can work together for the health of people and the chimpanzees, the human closest relative. We created a bond and I gradually became interested in conservation. Jane was determined and she had a vision of what she was doing, research and later community conservation and Roots and Shoots and I was inspired by her stories and talks even before we started discussing the possibility of me joining the conservation project.

I was captivated by her ability to maintain a low profile, being engaged in what is happening around the environment, people and wildlife, her friendly attitude, and her heart-touching stories, sense of concern about the levels of poverty and diseases. She always advocated the mission of conservation whenever she had an opportunity to hold meetings and gatherings. All of this made me risk leaving my permanent job with the government as a young practitioner to join the TACARE project that still had 18 months to close. Though I had supported the introduction of the TACARE project in villages it took me more than a year to decide on whether I should leave my job and join the project. Jane was there to support my decision-making through encouragement. She was determined to have an impactful holistic project that would serve the local people, wildlife and environment.

I finally joined the project since 1996. My joining Jane Goodall Institute and working with Jane and the team was a paradigm shift and it has been a learning curve for me and has completely changed the way I see the connection between people, the environment and wildlife. I was made to become a natural conservationist. I have learned a lot and I am proud to be part of change-makers.

Supporting community-based conservation:

Jane had wished to establish a project that would foster conservation efforts by involving the local communities under the government's leadership so for me joining the project was one of unexpected outcomes. Least to say, for me joining the TACARE project was based on mutual trust and encouragement from Jane and George. I was finally interested in taking on new challenges under the new project. Both of them provided a clear understanding of the project and encouragement for its future.

Definitely, implementing a community-based project needed a number of things including Jane's vision and trust, meeting the needs of the local people, fundraising to keep the project running and most importantly patience. We had successes and challenges; ups and downs and it was always encouraging to have Jane by our side no matter the prevailing situation. There were a few times when the project funding stopped, and Jane had to contribute her personal money and continue to mobilize funds to keep the project running. This is the hidden secret behind our success. She has always been the shoulder for us to lean on.

To you Jane, I say thank you for initiating the project and supporting all the efforts that changed the lives of the disadvantaged human and chimpanzee communities in Tanzania.

EMMANUEL MTITI – Director for Programs and Policy, JGI Tanzania

The Patience of Goodall

— Erika Fleury —

The first time I met Dr. Jane Goodall, it was as a fan. I was 28 years old and in the middle of writing a book about the history of primate rights. I had read countless pages by and about Dr. Goodall that inspired me, and I was excited to learn that she had a public appearance nearby. While waiting in a long line one summer day in Rhode Island, I was excited and nervous, incredulous that I could finally be face-to-face with one of my idols. Yet as the hours crept on, rumors swirled there may not be enough time for everyone to get a moment with her. Despite the uncertainty, we persevered, out of hope – and thankfully, so did she. Dr. Goodall stayed hours longer than scheduled that day so that she could meet with everyone awaiting time with her. She signed my tattered copy of *In the Shadow of Man* and we posed for a photo, the heat of the day showing on us both though I'm grinning widely.

Later, as I advanced in my career as a primate advocate, I would see Dr. Goodall at primatology conferences and events. Preternaturally peaceful and calm, she seemed otherworldly each time – never rushed, haughty, or what one might except of an international celebrity of whom everyone has demands and expectations. When she visits my colleagues at primate sanctuaries, she is known for always being generous with her time as she meets with caregivers and volunteers, and she always says hello to the chimpanzees, too.

Dr. Goodall's composure is easy to recognize but less simple to implement. I recently oversaw the emergency rescue of chimpanzees from a refuge that could no longer care for them. It took over three years of navigating complexities of the COVID-19 pandemic, and required raising millions of dollars to safely rehome 40 chimps in need. The chimpanzees' helplessness in such a terrible situation was paralyzing – and I feared failure. However, if we didn't rescue them, nobody else could... so we persevered. When the final ones were finally safely rehomed, Dr. Goodall sent a note of congratulations.

It takes wisdom to be patient. It takes courage to persevere. Like young Jane when she first stepped into Gombe to observe chimpanzees, when facing an unclear and uncertain future one must focus on putting one foot in front of the other to do what needs to be done. I have learned from Dr. Goodall the value of such patience, and how it brings about the most good to all.

ERIKA FLEURY – Author, primate advocate, and the Program Director of the North American Primate Sanctuary Alliance. She lives in Los Angeles and is still excited to see Jane Goodall every chance she gets.

erikafleury.com.

Memories of Jane in China

— Erika Helms —

D r. Jane in China – what adventures you had! Greg with his Jeep as chauffeur, cameraman, and good-humoured trouble-maker drove us to the airport for a flight to Chengdu, and people recognised you in the domestic terminal at the airport where you autographed boarding passes. When we arrived in Chengdu, do you remember the funny conversation when the newest young staff member introduced her boyfriend who said (in English) "I'm so lucky, she lets me sleep with her!" or something like that, completely out of the blue?! In Chengdu, we went to Buddhist temples for special vegetarian banquets, had events at the international school, and we visited the rural village where Zhang Zhe had spent time on her project.

And don't forget being driven as well in Michael's tiny green car that he somehow folded his tall frame into. It was a rather messy car inside and he had written messages to himself on dashboard in permanent marker. Michael provided impromptu tours of Beijing sights for you in that car, adding to the fun in Beijing.

The staff in the offices in China work so hard and they have treasured their time with you so much; you mean a great deal to them! Your renown in China is unique and it has facilitated you and the R&S programme to reach millions and have a huge impact. It's been my huge privilege to be a part of it. I send my love to you on your 90th, knowing my love is shared by so many in China and around the world. Happy birthday!

ERIKA HELMS – First Executive Director of JGI China/R&S Beijing, and now Global Manager of the Jane Goodall Institute Global

thejanegoodallinstitute.com

A Couple of Hours a Month They Said

– Evelyn Deiner and Pauline Stuart –

Pauline grew up in Australia and has followed Jane's journey ever since she went to Gombe in 1960 to study chimps. Pauline, much to the amusement of her family and other naysayers, was intent on travelling the world and seeing with her own eyes all the wonderful sights, animals, and other interesting pastimes, that she had read about in travelogues, National Geographic and other magazines, films and television documentaries. Pauline travelled extensively both before and after taking up residence in South Africa where she worked Monday to Friday as an accountant and as a volunteer for various conservation bodies during weekends. In 2003, Pauline joined JGI SA as a volunteer and within days was appointed as the bookkeeper, but little did she know that her offer to help with anything would have her multitasking capabilities tested to the extreme!

Evelyn joined JGI as a trustee in 2007. She was asked if she could provide some assistance on the legal side of things. Unbeknown to her, a slew of legal and other challenges awaited on the horizon, none of which were within JGI SA's control.

Roll forward to 2023, Pauline (though supposedly trying to retire) is still the accountant but has also taken on the roles of curio shop buyer, acting executive officer, general manager, and the all-around task-juggler – not to forget being the living archive and factual storyteller of JGI SA. Evelyn is the Chair of the Board of Trustees, which appears to be more like a full-time position.

Over the years, Pauline and Evelyn have stood together through many, many challenges, with some of them being heart stopping, "what now?"

moments. Through them all, it's been a comfort and privilege to know Jane and to have her unwavering support and friendship. Through all these trials, Jane has been a pillar of strength, always finding a reason for hope and providing inspiration to carry on.

Here's to 90 years of inspiration, hope and making a difference! Happy Birthday Jane.

EVELYN DEINER AND PAULINE STUART – Board Chair of JGI South Africa (Evelyn) and Sanctuary Manager at Chimp Eden, South Africa (Pauline)

chimpeden.com

Jane in Spain

— Federico Bogdanowicz —

When I saw Jane for the first time in 2007 in a conference in Barcelona, I was so humbled and inspired by this amazing 73-year-old activist that I decided to become a JGI volunteer right there. Now reaching 90, Jane keeps working tirelessly to make this world a better place.

Through all these years, as the director of JGI Spain I've had the immense fortune of sharing quality time and many anecdotes with Jane during her travels in Spain, Portugal, and even in Argentina. In between very important events, I remember Jane teaching me about the nidification places of sparrows in Madrid's airport, as we waited for the next flight; or suddenly challenging me to a race in a hotel corridor in Alicante (which she won), or leading by the example by taking the stairs instead of the lift every time she could (even with authorities), avoiding plastic bags, or saving the remaining half of the sugar bag for her next coffee. Jane is very austere with herself and very generous with others. She only buys clothes in extreme cases, for instance when she forgets trousers or shoes in a hotel. Once she wore only her black socks and black trousers to a very formal gala in France, as she had forgotten her shoes in the previous country. Nobody seemed to notice or care.

One of the few times she allowed me to carry her suitcase was when she came from Africa with a broken arm and cheekbone, after a 60-kg rock fell on her when climbing up a mountain in Gombe at around 80. On tour in Portugal, I had the huge privilege (and even bigger responsibility) of having to trim Jane's iconic ponytail, which had grown too long and did not resemble her famous profile in our logo. Along with English colleague Tara

Golshan, we managed to trim it back into the right shape, I believe (Note: the JGI logo was changed afterwards, but for other reasons, I was told).

Jane has made a huge difference in Spain, addressing both global and local issues in conferences, interviews and meetings with authorities. Apart from speaking of the climate and biodiversity crises years before the issues were on the national public agenda, Jane has advocated in defence of animals kept captive for entertainment (such as dolphins), or mistreated in public events (such as horses), or abandoned and euthanised at shelters (such as dogs), or exploited for intensive farming, or cruelly tortured and killed for "fun" (like bulls), or cowardly shot for trophy-hunting (like the elephant killed by the Spanish king). Jane has also defended natural spaces and endemic species in Spain (such as the lynx), advocated against octopus-farming, and endorsed the 3 chimpanzee rehabilitation centres in the country. Jane's support and guidance has allowed JGI Spain and its R&S program to grow in incredible ways. Jane's legacy will live on in Spain and Senegal.

Jane has had an enormous impact in many people's lives, including mine, of course. Apart from a mentor and friend, Jane has become the godmother of my baby girl, Amélie Jane, who will always carry a bit of Jane and her message of hope throughout life.

FEDERICO BOGDANOWICZ – Executive director of the Jane Goodall Institute Spain / Senegal

My Dearest Sister Jane

– Fred Matser –

It has been such a Joy to know and be befriended with you for several decades now. You are an inspiring and loving spirit to me and I feel blessed to be part of your life. We shared many joyful experiences and I hope it will be given to us to continue life together to share joyful, as well as sad times. I often think of your lovely mom Vanne.

Your life theme to give animals names was a breakthrough happening in the scientific world. As many of us know, alas not all, this has helped to give animals a 'face'.

Although we have not met that much in recent years I hope in the years to come that we may share joyful, happy and meaningful events.

I love you a lot.

Miles of smiles, sprinkles of twinkles, lots of love,
Your brother,
Fred

FRED MATSER – Founder of the Fred Foundation

fredfoundation.org/en/

Tanzanian Tribute

*— Freddy Kimaro, Deus Mjungu, Esther Sabuni,
Japhet Jonas Mwanang'ombe, Erasto Njavike,
and Anthony Collins —*

As we celebrate your remarkable journey we thank you for your dream to come to Africa, and we congratulate and wish you a happy celebration of your years. We imagine the bright light we would see from a birthday cake with 90 candles, shining on your face and the faces of your family... !

You have taught us so much about our primate cousins, giving us a window on all life and our place in it. We have learned so much from your stories, you help us see the world in a unique way, 'when we understand, we care...' We are so privileged to put your ideas into practice, and to share them with the rest of the world.

Starting at Gombe, we thank you because since that first day you stepped ashore, prior to your ground-breaking discoveries about chimpanzees, you have welcomed us to make further landmark studies together. And your benign attention has definitely secured the future of Gombe's chimps... your friend Flo is survived by 14 matrilineal descendants... ! If you had not come, would they be alive today?

But apart from the changes in Gombe, you also saw the changes outside, and that these changes came at a price: the over-exploitation of natural resources. You addressed this around Gombe by involving the community, and creating TACARE. But you also saw that many other countries suffer these same problems, so that Tanzania's TACARE now provides the world

Jane and Anthony Collins in Gombe, 2018.
Image Source: Thomas D. Mangelsen

a model to help save this planet for our future generations.

And you also recognised the power and adaptability of youth, so we also thank you for gathering twelve young Tanzanians to your home, and there creating Roots & Shoots; encouraging our young people to take action to make their country a better place for all living things, and thereby providing a model to assist other countries across the world.

In all, you have inspired so many of us that the impact of your life is more than 90 years! Every year of your work is multiplied-up by the number of people you have inspired along the way, (such as our current 522,173 members of R&S!), every year new companions join, you have become year-by-year more effective. So your impact is already many times more than 90 years-worth, you have already lived many lives, to such great effect.

You also mean so much to us personally, for what you have given to our own lives. Your enthusiasm with saving tomorrow's world for our children inspires us all... "If I am asked, how should I call her, its simple, she is a walking angel." And while your hours and days are fully committed to your mission, yet we know you are a compassionate caring and loving person, and we are lucky to be part of your larger family, we count you as Our Mama.

Our concern now, although you have done such a lot, is that we still owe you a lot: we have so much yet to do. But the one thing that gives us hope and courage, is your reassurance that we are not too late, there is still time for us to make this world a better place for everyone, and we will do so. "I promise to live your instructions always, I will do everything I can to live and protect your philosophies, visions, dreams, and ideas."

And in a lighter vein, just like vintage wine, some things get better with age! So here is to another year of good health and great times. Happy 90th birthday, and we will see you at your 100th!

Freddy Kimaro (Executive Director JGI Tanzania), **Deus Mjungu** (Director Gombe Stream Research Center), **Esther Sabuni** (Chimpanzee Research), **Japhet Jonas Mwanang'ombe** (Director R&S Tanzania), **Erasto Njavike** (R&S Manager, Northern Zone) **and Anthony Collins** (Director Baboon Research)

The Caring Woman

– Galitt Kenan –

Dear Jane,

The first time I met you was in Paris in 2008. You were with your friend Yann Arthus-Bertrand. And at the time, I knew you "only" as a public figure. I knew that you were a role model, a leading scientist, on the field (I thought you were an ethnologist, not an ethologist!) and that you had set up a well-known conservation NGO. For me, you were the woman who protected chimpanzees thanks to the strength of your will, your work and the many activists that you inspired.

I didn't realize that every time you speak publicly, tears flow, eyes sparkle and hope rises. That the standing ovations can last up to two minutes (two minutes is a long time!).

The years went by and I started reading, one, ten, fifteen of your books. And each time I fell under your spell. I discovered the passionate activist, the woman of Peace, who listens and respects Others (humans and non-humans). Each time, I closed the book with the desire to act, the desire to be part of your team. I was lucky enough to join "Team Jane" in 2018. Since then, I continue to be even more impressed by your personality, by everything that you are, and that you do.

I've since been privileged to discover the woman behind the icon. The caring woman who checks up on you when your child is ill and gives you advice and support when things are difficult. The woman who wants to know how you're living, whether your life is balanced and happy. The woman who takes the time to know all the extraordinary things that the wonderful volunteers are doing in France, and thanks them all, one by one. The woman who tries to get you to break your bad habits (Jane, I'm afraid it'll be many decades before I stop apologizing all the time!). The woman with whom you can talk about the thousand and one ways to change the world, in the evening, with a drink, on the floor of a hotel room. With whom you can talk about extremism, secularism, veganism, fascism, the birds in the tree, the books that have left their mark on you and your everyday life. And to laugh with.

The woman who gives you hope. And the desire to do more.

Always and again.

To everyone, and to me in particular.

Thank you Jane and happy birthday!

GALITT KENAN – Executive Director JGI France

Happy 90th Birthday, Jane

– George Schaller –

I thank you for the honor of asking me to contribute to your book celebrating Jane's 90th birthday. I have tremendous admiration for Jane's dedication to wildlife and their habitat over the past decades, and communicating this to the public in various ways. I first met her in 1960 during a visit

Jane in her tent during the early years in Gombe.

Image Source: Hugo van Lawick

to Gombe at the start of her long-term ground-breaking research. I wish Jane at least another decade of fieldwork and of entrancing her audiences in auditoriums.

GEORGE SCHALLER – World renowned zoologist, ecologist, and conservationist

Jane and Jumanne Kikwale, 2019. Jumanne worked for many years as people and camp manager at Gombe. He was seven years old when Jane arrived in Gombe on 14 July 1960. He lived in Gombe as the son of Rashidi Kikwale, Jane's main assistant during the early Gombe years.

Image Source: Anthony Collins

The Fauna Family and Jane

– Gloria Grow and Tony Smith –

Turning 90, what a blessing, and a what a nice opportunity to thank you for who you are !

Almost 26 years ago, the Fauna Foundation was established, primarily as a refuge for ex-biomedical chimpanzees from a laboratory in upstate New York. In 1997, we welcomed 15 residents over the first few months. Others joined the Fauna family later, having been rescued from three different zoos in Quebec. We also welcomed two unique signing chimpanzees from Washington university's CHCI facility, famous for its ASL (note from the Editors: American Sign Language) chimpanzee research. In total, this amounted to 23 wonderful chimpanzees and 7 amazing monkeys. We take great pride in this endeavor, and, though it has had its ups and downs, it's something we have never regretted being a part of. We continually strive to make improvements and be worthy of the trust of our residents, who we consider our precious family and whose stories deserve to be heard by humanity.

Jane was a guiding star for us during this process. She provided so much positive encouragement, advice, and feedback, which has been rooted in our hearts; the rest is history. Jane's support even continued beyond that point when she accepted our invitation to join the Fauna Advisory Board. She truly helped us to survive and thrive as a sanctuary. For that, we are eternally grateful.

Much to the delight of both chimpanzees and human caregivers, Jane has made time out of her busy schedule to visit Fauna on several occasions. After each, we are always left with a sense of renewed motivation and purpose. Her interactions with the chimps are so precious to them. You can clearly tell that they recognize her as a kindred spirit. Additionally, she has always paid special attention to the children of Fauna. Many of our younger family members from Tony's daughter, Maggie, in 2001 to our grandniece, Poppy, in 2023 have come away feeling seen and recognized, which is so crucial at that age.

Jane is a strong and relentless woman who is determined to empower those around her. She provides the lives she touches with inspiration and a sense of responsibility – not only for their fellow humans, but their fellow animals and planet. Though Jane's achievements are far too many to mention, something that strikes us, at Fauna, is her way of reaching a global audience to deliver her message of never-ending hope. As Jane approaches her ninetieth birthday, she continues to travel the globe with the same amount of energy she had when she was observing chimpanzees in Gombe many years ago. We have had the privilege of witnessing her extraordinary will to create change while respecting all creatures, human and non-human.

Over the years, we have had to contend with the passing of many members of our family, majorly from the consequences of the intrusive protocols that they were subjected to during research. Each loss has had a devastating effect on the Fauna family, which includes our wonderful caregiving staff, volunteers and donors.

We remain committed to providing the loving care necessary for the six remaining chimpanzees and lone monkey. Jane remains an important figure to us during these times because she reminds us that nothing is impossible.

We all wish Jane a happy ninetieth birthday and look forward to seeing her in the future.

GLORIA GROW – co-founded the Fauna Foundation with her partner Dr Richard Allan, the first and only chimpanzee sanctuary in Canada.

TONY SMITH – member of Fauna's advisory board and dear friend and brother-in-law to Gloria Grow.

faunafoundation.org

– Grub van Lawick –

I am writing this in Gombe where I've come for a short five-day trip along with my mother, daughter, and eldest son. Today is our last day together.

We are staying at my mother's old house on the beach, overlooking Lake Tanganyika on the one side and the thick forests of Gombe on the other.

Every time I come here it whispers old memories, like a pathway through the mists of time.

An old faded handwritten note stuck to the entrance of my mother's bedroom says in Swahili, "This is the room of Jezebel the spider, she must not be disturbed".

When first posted it was meant as a simple reminder to the cleaning lady, that the very large poisonous looking spider, whose formidable web had graced the space above her mosquito net, must not be harmed!

Jezebel has long since passed but as her various offspring enjoy equal privileges, today her lineage is going strong (I try to avoid that room).

Yet, if one was to assume that such privileges are extended to all members of the animal kingdom by her, then one would be most mistaken, for I remember a time when my mother declared a war on rats!

I believe the final straw for these unfortunate creatures came after they attempted to turn some of her most important research papers into a fine dining experience.

With a determination that always follows any decision she makes; she embarked on a mission to permanently expel every member of the rodent clan from her Gombe residence. However, if one was to assume that this plan would entail traps of the guillotine kind, then one would be equally mistaken. For not only did she insist on live traps, but these were to be specially designed, to ensure that no part of a rat's anatomy would be harmed during the trapping. And so it was that these live traps, designed with the utmost care for the rats' well-being, were made and discreetly placed throughout the house.

These devises proved to be highly effective and so it was that the evictions started. Every morning our first job would be to carefully place all trapped rats into a large bucket, after which we'd go for a long walk along the beach, releasing them into the forest, once she believed that we were far enough away that they wouldn't return.

Today in her twilight years her compassion for animals has only grown stronger. A recent incident from three days ago, showcases this only too well.

Whilst my mother was on her morning walk along the beach and my son was out hiking in the forest, I was enjoying a much more relaxed morning with my daughter Angel at the house, reading a book, when suddenly with dismay she noticed a trail of biting ants leading all the way from the front door to the top of her rucksack. The discovery was soon made that a forgotten packet of crisps within her bag was the source of their attention.

Jane reading in her home in Gombe, 2018.

Image Source: Thomas D. Mangelsen

We now had the challenging task of extracting Angel's belongings from a swarm of biting ants, visibly agitated by our interference. I am ashamed to say that during this tricky endeavour neither of us for a moment gave any thought to the welfare of said ants.

In stark contrast a short time later my mother, returning from her walk, enters and upon seeing this seething mass of biting creatures on the floor, her first words were. "Poor ants what's disturbed them?"

It was obvious that whilst we saw a nasty pile of vicious pests, she saw a mass of displaced living things that were in obvious distress. That whilst individually small, in her mind, each one of them was a tiny speck of consciousness, fellow travellers on this journey that we call life and thus deserving of her protection.

Her desire to right this obvious wrong went as far as to suggest that we should return to them the packet of crisps, that we had subsequently chucked in the bin, so that the ants might be allowed to finish their meal and thus recover from being so unceremoniously dumped onto the floor!

As neither myself nor Angel had any desire to remove these from our dirty compost bin a compromise was reached, whereby two crisps from a new packet were placed in the corner of the room. This worked a treat for a half hour later we were all happily enjoying our lunch together.

Us with our plates of vegetables on one side of the room and the migrated ants on the other, with their two large crisps!

No doubt the Jezebels were also feasting in the next-door room, but as I had no desire to check upon their wellbeing, I guess we shall never know!

GRUB VAN LAWICK –Jane's son

We'll Meet Again, Sweetheart

– Gudrun Schindler-Rainbauer, Diana Leizinger, Doris Dienst-Schreyvogel, and Walter Inmann –

Jane's connection with Austria is a decades-long one. Even before the Institute was founded in 2003, there were several friendships and connections with this small country. But the great bond began when Walter Inmann and Melissa Tauber laid the foundation stones for JGI Austria at that time. Year after year there was a reunion with Jane. Every visit means a huge mountain of tasks. We appreciate and admire you immensely, Jane! And we try to support you whenever we can. But it is not the work alone that binds us together. It is rather those many happy hours, moments of fun, the intense conversations and often curious experiences beyond duties. Do you remember your complicated journey to Austria, your landing in Salzburg without luggage? When you got a bag of clothes after your lecture? Forgotten things from a cleaning shop? It was the beginning of our friendship with one of our ambassadors of honour. We remember the evening on the shore of Lake Traunsee: we toasted with whiskey – you told us to take the glasses from the bathroom of our hotel. There are many anecdotes. There were many long evenings with you and your friends from Austria: whether in the British Embassy in Antonio's kitchen, whether in Walter's mountain hut in Reichenau, whether at our homes or in the flat of Ulli Goldschmid, whether at the Heuriger. We always discuss the state of the world, tell each other gossip and from time to time you make us laugh with your jokes. You tell us about the big JGI family and stories from all over the world. You share some of your worries with us. We are always there for you and we also accompany

Sonja Aichinger, Doris Dienst-Schreyvogel, Diana Leizinger, Jane Goodall,
Gudrun Schindler-Rainbauer, Nikola Reiner-Rautek and – in front – Walter Inmann
in the garden of the British Embassy of Vienna, 2015.

Image Source: The Jane Goodall Institute Austria

you on tours and some trips: Whether Hungary, Italy, Germany, and Switzerland, even in Africa we are by your side. We pack fruit and leftover food for later moments. Rescue you from situations when something is bugging you. Take care of you, Mr H, Cow and your seven animals. There are so many things we associate with you. But never will we forget the evening in Budapest when we first sang the song, made famous by Vera Lynn, *We'll Meet Again, Sweetheart.*

You can always count on us, Gudrun, Diana, Doris, Walter and the team from JGI Austria! As it is natural for a family. We love you and wish you all the best from the bottom of our hearts!

Gudrun Schindler-Rainbauer, Diana Leizinger, Doris Schreyvogel, and Walter Inmann – Jane Goodall Institute Austria

janegoodall.at

A Birthday Note

— Ian C. Gilby —

Dear Jane,

Happy 90th Birthday! What an incredible journey you're on.

Hosting you at the Gombe Chimpanzee Research Archive recently was truly one of the highlights of my life and career. It was clear that reading your original notes transported you, and by extension, the rest of our research team, back to Gombe. For you, exploring the archives must have been like

Each evening in her tent at Gombe, Jane would enter data from her field notebooks into a journal.

Image Source: Hugo van Lawick

revisiting old friends. Hearing your stories about the early days at Gombe first hand will forever give me goosebumps, and I know that the time you spent with the students here changed their lives. You have a unique gift for inspiring others to follow their passions, and to do good in a world where it is so desperately needed.

And you refuse to slow down! The new partnership between Arizona State University and the Jane Goodall Institute marks another major step toward healing our ailing planet. In the face of monumental environmental challenges, our combined efforts promise to unlock even more of our world's mysteries by highlighting the interconnectedness of all its inhabitants. It is truly an honor for all of us to work alongside you. Your humility, wisdom, and willingness to embrace fresh perspectives are a testament to your dedication to the betterment of our world. Your passion for conservation and your advocacy for change at the local level remind us that the pursuit of knowledge should always be in service of something greater – the protection of our environment and all the creatures that call it home.

As you celebrate this milestone, please know that your impact resonates far beyond the scientific community. Your legacy is woven into the fabric of human history, and inspires so many to look at the world with empathy, curiosity, responsibility, and of course, HOPE.

So, here's to you, Jane! May your 90th year be as impactful as the stories you share with the world.

Love,
Ian

IAN C. GILBY – Associate Professor, Arizona State University

Speaking Out for Others

— Ingrid Newkirk —

Turning 90, what a blessing, and a what a nice opportunity to thank you for who you are !

In 1986, while Jane was staying with longtime PETA member and author Ann Cottrell Free, "the case that broke the species barrier" erupted.

A group called True Friends had somehow surreptitiously entered SEMA, a National Institutes of Health–funded laboratory in Rockville, Maryland, not far from where Jane was staying. Once inside, they made off with four baby chimpanzees slated to be used in hepatitis B and AIDS experiments.

Group members filmed conditions inside the laboratory, where adult chimpanzees were locked in solid-walled steel chambers barely larger than their own bodies. They rocked back and forth, banging their heads against the cage walls, with nothing to see but a few feet of the room and nothing to hear but the constant hum of the mechanism pumping air in and out of their chambers. The infants had yet to be infected when they were spirited away into the night, never to be found.

PETA showed the video taken inside SEMA to Jane, who watched it with "shock, anger, and anguish" and decided she must go to the laboratory to see inside it for herself. Government inspectors had been turned away, but how could they turn away the world's foremost chimpanzee researcher? They refused Jane at first, then realized they had to allow her, accompanied by Sen. John Melcher of Montana, to enter. Both of them emerged shaken.

Jane wrote, "I shall be haunted forever by ... the eyes of the infant chimpanzees I saw that day," and went on to describe the cages as "bleak and sterile, with bars above, bars below, bars on every side." She made a pledge: "The least I can do is to speak out for the hundreds of chimpanzees who, right now, sit hunched, miserable and without hope, staring out with dead eyes from their metal prisons." And she did just that.

SEMA changed its now publicly tainted name to Bioqual, but more than that changed as a result of Jane's involvement. Chimpanzee babies held in the facility started to be housed together and be given toys to tear apart, some tiny compensation for being deprived of a real life and eventually infected with diseases. But eventually, this case became a catalyst for the biggest change of all: the end of the use of chimpanzees in experiments in the U.S. PETA thanks Jane for all she has done to save animals.

INGRID NEWKIRK – Founder of PETA.

peta.org

On the Road to Nazareth

— Itai Roffman —

Turning 90, what a blessing, and a what a nice opportunity to thank you for who you are !

I wish you 120 years of good health, great success on all your efforts for awakening humanity to save our earths endangered ecosystems and may your dreams come true. Amen. May the Great Spirit protect you always.

As our lives were intertwined for a big part of my life, it is difficult to think what I can share with you that you don't already know about me. Of course you are my inspiration in everything I do, your moto for me of "Never take no as an answer" and "Never give up, seize opportunities as they may not return again" are manifested on my journey. You have encouraged me in every major step in my life to achieve success, first I needed to get my high school diploma, then Bachelors, then Masters, and finally PhD (your letters of recommendation allowed me to get scholarships for these); you remember telling me that no one would listen to me without a Dr. before my name.

Jane, you took me out of the wetland with my small butterfly net trying to protect the Syrian Cat Eyed Spade Footed Frog (funny name for an endangered species you said) and sent me to the field in Mali to document and protect the cultural diversity of the critically endangered Sahel cliff-dwelling arid Savannah woodland chimpanzees. Your UN Messenger of Peace title inspired me to create peace accords between the tribal chiefs and the chimpanzees (just like when we flew your beautiful giant peace dove across the three borders of Israel, Syria and Lebanon with the Alawite R&S Rajar Village kids). As you know, because you wonderfully guided me to apply for the grants that supported it, our Mali partners and I signed 32 tribal village peace accords and managed to protect 2,500 km^2 of chimpanzee cliff-range habitat. We thanked the tribes with very needed aid enabling them to establish nurseries for their ancient medicinal plants to ameliorate Malaria and other ailments, and grow endemic trees for ecological restoration. In the past 13 years since the Mali program began three wars came with the 3 coup d'etats but as your perseverance taught me that nothing will stop us from continuing our conservation and R&S Mali efforts especially if our intentions are righteous. We established the Savannah woodland sanctuary at the National Zoo of Mali and gave a better life to those amazing 3 captive orphan adult chimpanzees. They thank you, I thank you.

I know how spiritual you are, as I am, since we were both adopted by Native American tribes, so I must tell you that much of my research and conservation guidance came from the spirits of chimpanzees who died at the hands of man and Native American grandmothers who suffered the most tragic fate, I wanted to take this opportunity to acknowledge them.

Well, finally, we have more projects to fulfill on our journey together, and your establishing with me of JGI Israel at the Max Stern Yezreel Valley College (downhill from Nazareth and across the road from the historic mound of Armageddon) will enable us to make our vision reality.

Our next R&S mission is rewilding from zoos to nature in Israel, Palestinian territories and the Arabian Peninsula countries surrounding us with extirpated critically endangered species from the Addax, Wild Ass, Syrian Brown Bear, Arabian Gazelle, Nubian Ibex, Arabian Leopard, Golden Squirrels, Persian Fallow Deer and maybe we could convince Saudi Arabia to bring back their Asiatic Lions extinct for over 120 years there.

Happy Birthday Jane!

May personhood rights for nature as envisioned by indigenous sovereign nations be implemented across the world (including endangered ecosystems/wildlife and of course our Hominid/Hominin Great Ape family).

Safe travels and many more years of great projects, accomplishments and discoveries to you!

Lots of love forever.

ITAI ROFFMAN – Co-executive Director, the Jane Goodall Institute Israel, Researcher/Senior lecturer, the Max Stern Yezreel Valley College

— The Jackson Hole Women: Penny Maldonado, Sue Cedarholm and Tiffany Talbott —

The Jackson Hole Women are Penny Maldonado, Sue Cedarholm, and Tiffany Talbott. They're close friends of wildlife photographer Tom Mangelsen and work closely with him.

The Heartbeat of the Planet

Thinking of Jane brings a gallery of cherished memories to mind. These images are snapshots of sheer joy, captured on the faces of those who hear her name, and the recollections they hold dear. Her influence radiates happiness across borders and perhaps even species.

While Jane's superpowers are infinite, flourishing even before that term became popular, the three that seem to fit into everything she does are commitment, compassion and connection.

Commitment, best illustrated by the wonderful idea that the only limits are those of vision, grew from her early curiosity into where a chicken's egg came from to her lifelong dedication to exploration and inquiry, pursued with a gentle determination to this day.

Compassion, interwoven with scientific inquiry, flows through her work. While Jane is a scientist by trade, she defies the cold constraints of data. She unveils the realm of animals as sentient beings – emotionally rich, intelligent, and endowed with unique identities. Jane frees us to give them names! Through her, we learn not just to care through knowledge, but to bond instinctively with all species, acknowledging our role as human animals on this planet. Her commitment to compassion nurtures a web of understanding that unites us all.

Jane's unparalleled gift for forming connections stands as a testament to her spirit. Who can forget the brave, young, Brit, sitting waiting for the first connection with the chimpanzees? Who has not been moved to tears by the hug of Wounda as she thanks Jane at the start of her liberation journey? Whose heart doesn't swell when children from every corner of the globe proudly discuss their Roots and Shoots projects? Jane bridges chasms that some might perceive as barriers, crafting them into thrilling opportunities. It is diplomacy without language, love without borders, adventure without limitation.

In her recent talks, Jane delves into the very final frontier – the ultimate adventure – one that unites us all in its certainty. With fearlessness and compassion, she approaches the greatest unknown, recognizing it as life's inevitable commitment. This unwavering stance fosters a connection that resonates deeply within us all.

And back to the joy, her name lights up faces, her kindness has changed and indeed saved lives. Jane's tenderness is her strength and steel. She is indeed the heartbeat of the planet.

PENNY MALDONADO – Executive Director, The Cougar Fund

A Learning Moment with Jane

Jane was in Jackson, Wyoming giving a talk at the Wildlife Film Festival. After her talk Susana Name, Jane's assistant, and I were helping her get through the crowd. A young girl came up to Jane and wanted to shake her hand, Jane stopped, shook hands, and talked to her for a few moments. Later, when we were all in a room in the hotel Jane told us how she always stops and takes time to meet people, especially young people, because you never know who may be the next great conservationist. That has always stuck with me, we never know how we can perhaps influence or open the eyes of another. It has been an honor to be able to have spent time with Jane over the years. Happy Ninety Years!

SUE CEDARHOLM– Artist, photographer, and personal assistant to Tom Mangelsen

Behind Every Great Woman is Another Great Woman

"She supported me from the very start," Jane has said of her mother, Margaret "Vanne" Myfanwe Joseph Goodall (1906 -2000), who was an incredible woman in her own right.

Jane's dream as a young child was, "Grow up, go to Africa, live with wild animals and write books about them." Many parents in the 1930s would throw sand on such a fantastic dream. Jane said, "Everybody laughed. Except

Jane and her mother Vanne sort specimens in Gombe.

Image Source:
Hugo van Lawick

mum." Instead, Jane's mother told her, "If you really want to do something like this, you'll have to work really hard, take advantage of every opportunity, and never give up."

Jane recalls in her talks how she disappeared as a small child to observe how a hen lays an egg. When her frantic mother found her, rather than chiding Jane, Vanne sat down and listened to Jane tell of what she had learned. Jane says, "Isn't that the making of a little scientist? The curiosity, asking questions, not getting the answer, deciding to find out for yourself, making a mistake, not giving up, and learning patience. A different kind of mother might have crushed that spirit of curiosity and I might not be standing here now."

Thank you Jane for being a pioneer in science, for working tirelessly for chimpanzees, animals and peoples around the world. Thank you Vanne for leading the way, for raising and nurturing the incredible spirit of your daughter.

TIFFANY TALBOTT – Photographer and animal welfare advocate

Jane the Storyteller

— Jae Choe —

There is an undeniable pathos in the voice of Jane Goodall. Anyone who has ever served in her entourage knows how little she eats and that her voice is feeble and rather monotonic. And yet I have witnessed time and again that her speeches bring tears to the eyes of the crowd. She hides an iron fist in a velvet glove and her voice is no exception. Jane is certainly not a tempestuous orator but passionate storyteller. Her storytelling has a special power to captivate the hearts of people of all walks. Not only small children but also grey-haired adults are helplessly enthralled by her tales. I've been a designated translator since her visit to Korea in 2003. During her seven visits for the past 20 years, I must have translated her lectures well over a dozen times. Quite embarrassed to admit but I have wept at least three times while translating her speeches.

The most unforgettable was when she gave a talk at a women's organization during her visit in 2007. Jane was telling the story of a man's rescue of a drowning chimpanzee at Detroit Zoo. Pulling a chimpanzee out of the water is such a tough and heroic act because chimpanzees can't swim. It is also extremely dangerous because chimpanzees are much stronger than us and can be viciously brutal. When people asked the man how he could act so bravely, he answered plainly that he could not help but jumping into the water looking at the chimp's eyes begging for help. In the middle of her telling this tale I suddenly burst into tears and could not stop myself until Jane walked up to and hugged me around my shoulders. A middle-aged man sobbing like a little kid in front of several hundreds of women was a scene to remember.

Jane Goodall is a gifted storyteller who makes everyone 'root and shoot'. I wish that I could continue to hear her voice for another 90 years.

JAE CHOE – Distinguished Professor of EcoScience, Ewha Womans University, Founder and Chair of the Biodiversity Foundation

Jane Goodall Influencing Power

— James Lembeli —

I feel a proud and most fortunate man to have known a lady who has done lots for our planet earth. Her life has changed the lives of millions of people all over the world, people of many colors, of different age categories from young ones to elderly men and women. Jane's life has been inspirational to world leaders, wealthy people, scientists, and ordinary people. Her struggle for making the world a better place despite the ongoing biodiversity loss, deforestation trends, pollution and climate change threatening the inhabitants of planet earth. She never gives up her struggles despite her age. I am also amazed how Jane remains connected to her friends scattered all over the world, on different continents, people who help making the world a better place for humanity and nature.

I recall that before meeting Madam Jane I had an opportunity to see her amazing films on the Gombe chimpanzees. Most of them were made with the help of the National Geographic Society. I first met Jane Goodall in person a month after I got my employment with the Tanzania National Parks as a Public Relations Manager in 1994. When I visited Gombe with the Director General of Tanzania National Parks, Mr. Lota Melamari, we were lucky to find her in Gombe, we visited her cabin, she was babysitting an orphan baby chimpanzee. I asked my boss, what is she doing? He just laughed. Later, I also got an opportunity to visit her first husband Hugo van Lawick who spent the last years of his life in Kirawira within the Serengeti National Park. I was surprised with the couple because, though separated, they were still living in similar ways. Their lives were quite simple and showing satisfaction with fewer material possessions, but connected to nature. I continued

James Lembeli, Jane and Erasto Njavike in Arusha, 2019.
Image Source: Jane Goodall Institute Tanzania

following her project because I worked with the Tanzania National Parks, which controls all the national parks including Gombe National Park, then left the job for taking up a parliamentary position. My love for nature was growing to the extent that I was not shy to stand for nature whenever I realized that it was in jeopardy. One of my major concerns was when I stood up among the people who opposed the construction of a tarmac road within the Serengeti National Park, which was a politically sensitive issue.

Luckily, after my 10 years term in Parliament, the Jane Goodall Institute elected me to become the chair of the Tanzania JGI Board from 2018 to 2023. I was also assigned to become a JGI US Board member since 2018. During this time I had an opportunity to accompany Jane when I participated in various fora in Tanzania, USA, France and Germany. My time around her has continued shaping me and giving me some valuable qualities just by seeing a glimpse of her lifestyle. The truth is, with what Jane does one would wish she remains on planet earth forever, but I do not think this will happen. However, I am very glad that she is very busy making more Jane-like people all over the world. Just wishing her many more years on planet earth, so that she continues making more Jane like kids amongst the younger generation.

JAMES LEMBELI – Tanzanian politician, Chair Emeritus of the Board of Directors of JGI Tanzania and member of the Board of Directors of JGI USA

Dayenu: It Would Have Been Enough

— Jared Polis —

I first met Dr. Jane Goodall in the early 2000s, when I was in my twenties and serving on the Colorado State Board of Education. Riveted by her Roots & Shoots program, we explored how we could expand it in Colorado. But my awareness of Dr. Goodall dates back nearly two decades earlier to my younger sister's sixth grade biography fair, in which she portrayed a young Jane. In high school at the time, I marveled at the way my little sister found inspiration in Jane's life and work. My sister later went into science and received two Masters degrees in science, inspired in part by her study of Jane.

While plans to establish a Roots & Shoots-specific charter school never came to fruition, elements of the program expanded into a number of Colorado schools and also helped to further expeditionary learning. Colorado is a place where we pride ourselves on our relationship with nature, so it's no surprise that Jane's work helped influence generations of Coloradans with the values of sustainability and co-existence.

While serving in Congress from 2009-2019, I met Jane several times in Washington, D.C., and once even ran into her at an airport lounge in London. Her message of peace and hope helped inform my work in the United States Congress. It wasn't so much the science itself—though the special kinship between chimpanzees and humans made that work innately captivating—but rather, the broader message. *Life is precious*, and we must learn to get along with one another and with the planet to ensure our mutual survival and success. Jane stated that "the greatest danger to our future is apathy," and her words and work galvanize us to impassioned discussion and, more importantly, action. Whether the issue at hand was climate change, peace

Governor Polis, Jane, and Marc at a Roots & Shoots event
at the Mackintosh Academy, Boulder, Colorado.
Image Source: Susana Name

among nations, conservation, or education, I constantly found strength and wisdom from Dr. Jane Goodall.

Jane not only took her life's work, underscoring the plight of chimpanzees and the urgency of African conservation (as we say in the Jewish tradition, "Dayenu": *that would have been enough*), but she then successfully applied the power of the lessons learned in a manner that is vigorously relevant to our lives and the future of the planet. Very few scientists are able to popularize their own work, no less dynamically spread a powerful moral message. Jane continues to do both, all while serving as an ambassador for feminism and, indeed, science itself. It's remarkable that this amazing force of nature from Bournemouth, with genius forged in the wilds of Africa, has such a profound impact not only in our remote mountain state of Colorado, but across the entire world. Her message of peace and hope inspires us all every day to do better and to work harder to defeat apathy. As Jane herself stated, "Every day you live, you impact the planet; you have a choice to make an impact in a positive way."

JARED POLIS – Governor of Colorado

With Love, Dearest Jane, Always Your Samwise

— Jason Schoch —

Something Jane Goodall said to me in 2001, saved my life. It's because of the friendship that began that day and her mentorship of me as a human being, I was set on my life's path. Jane and I share a love of the works and words of J.R.R. Tolkien. When she asked me to return to the Northern Great Plains of the Dakotas to help her start her Roots & Shoots program in tribal communities on the Pine Ridge Reservation, I agreed. It's a tough place, broken by hard times. Yet we knew that the deep roots were not touched by the frost. Jane called a meeting in the Black Hills of South Dakota. Only a few braved the snow storm and a Fellowship was born.

Since then, Jane's quiet, enduring and indomitable spirit has guided me into a life of service to others. We began as humanitarians, as the persistent poverty, trauma and despair on the reservation, meant that's where we had to start. Today we're working to create biodiversity and wildlife-friendly, regenerative tribal farms on and around Pine Ridge that are guided by the traditional ecological knowledge and by the culture of the Oglala Lakota. Community-centered conservation, as practiced by Jane's institute's Tacare program in Africa has seen incredible success. When we began our work, we modeled it after Tacare in order to grow Roots & Shoots. Jane's innate understanding of how to reach people and to understand where they're at and what they need in order to gain their support of conservation efforts around Gombe has in every way, saved Gombe and the chimpanzees. I knew a similar approach would help us.

In 2017, whilst visiting with Jane, in her small forest home, hidden away in the forest of Gombe National Park, discussing my day's adventure to observe and photograph the chimps high up in the mountains, I watched as a tiny tree frog leaped from between the bars of the window down onto Jane's shoulder. She didn't startle, as most people would have. She just greeted it with a simple, "Hello" and kept on talking with Dr. Anthony Collins and I, sipping scotch. After some time, she asked me to help her move the frog back out into the forest. We stepped out into the cool darkness of the forest night, alive with the sounds of all the biodiversity protected by Jane's efforts and I helped her gently place the tiny tree frog on a branch of a nearby tree.

Returning home to the States, I knew we had to expand our efforts beyond helping just people, to start protecting the land, plants and animals around them. Industrialized agriculture is among the main drivers of climate change and biodiversity loss. It's ruining the physical and cultural landscapes of the Lakota. So we needed to add re-wilding and conservation efforts to our programs, in order to protect and restore the very place upon which the Lakota people lived and we started these efforts in 2018.

Note: I always sign off my letters and emails to her, "With love, Always your Samwise, Jason." Jane once called me her gardener and when I signed off a letter to her with Always your Samwise, she started to call me that.

JASON SCHOCH – Jane's Gardener. Born and raised on the Northern Great Plains in North Dakota and now residing in South Dakota, he has spent the last 18 years working on behalf of tribal youth and disabled people on the Pine Ridge Reservation helping to revamp the community food system to better meet the needs of the people, plants, place and the animals who all share those borders.

Jane Goodall: Our Planet's Leading Inspirational Force for Good

— Jeff Horowitz —

This essay was written to honor my dear friend Jane Goodall who could not have been a more wonderful partner in our collective fight to save forests and our climate!

Our beautiful Earth, a vibrant tapestry of blue waters, majestic mountains, and green forests, is under siege. Once home to six trillion trees, a mere three trillion remain, with half of that loss occurring in the last century. As we all know so well, the relentless pursuit of short-term profit continues to endanger the health of our planet.

But amid this disheartening reality, there's hope and action led by people, organizations, and governments realizing that the fate of forests is tied to the fate of humanity. Forests support incalculable numbers of life forms not to mention humans. And most urgent is the connection between saving forests and saving our climate! The forest-climate strategy is big, one-third of the solution and gaining traction as a cornerstone in our fight against climate change.

In this vital battle, no one shines more brightly than Jane Goodall! Her tireless dedication to protect, restore, and plant forests has reached the hearts of tens of millions around the globe. From the halls of the US Congress to the biggest stage at Davos, to the major UN international climate gatherings – when Jane speaks, people listen and act.

Her message, delivered with grace, humility, and humor, has mobilized young and old, veterans of environmental causes and those newly awakened to the urgent need to green our planet again. Through her words and actions, she has made the forest-climate strategy not just a solution but a rallying cry for humanity.

As we celebrate Jane Goodall's milestone 90th birthday, we find ourselves reflecting on the magnitude of her contributions. Her love for nature, her unwavering commitment to our planet's wellbeing, and her inspiring approach to activism have shaped not only the environmental movement but our very understanding of what it means to be a steward of the Earth.

In honoring Jane, we renew our pledge to her vision, recognizing that her life's work is a testament to the power of hope, action, and compassion. Happy 90th birthday my dear partner in peace. And thank you for being our guiding star in our collective fight to make our planet green again!

With much love on this special day!
Jeff H / PIP

JEFF HOROWITZ – Founder of Avoided Deforestation Partners

Note From the Editors: For a short film on saving our forests, see youtube.com/watch?v=RvdiVkNQD2s. It is narrated by Jane Goodall and produced, written and directed by Jeff Horowitz.

Our Friends the Chimpanzees

— Jenny Desmond —

Happy Birthday Dear Jane! What an honor to be included in your birthday celebration! Who would've imagined when my Mom gave me the book "My Friends the Wild Chimpanzees" at the age of seven with an inscription "to our very own Jane Goodall", that I would one day not only follow in your footsteps, but be graced with your love and support.

Of course it was a thrill to meet you for the first time, to spend time with you at events, and most recently, to be presented IFAW's animal rescue award by, of all people, THE Jane Goodall!

But the times with you that stand out most for me are far more simple.

Sitting on the patio of the chimp house in Uganda, drinking whiskey, and eating takeout from the local Indian restaurant.

Learning your sense of humor while taking picture after picture after picture.

Our beloved Princess giving you a giant kiss on the face.

Going to the Botanical Gardens to take a picture where they filmed Tarzan to your Jane.

Simple moments shared with a special person who brightens up every room, every conversation, every activity, with her presence.

More profoundly, while we sat talking about our other than human animal friends and family, it suddenly struck me that, while I knew many people who had an immense love and respect for all species, I recognized in you that innate feeling of sameness we'd both had from birth. Not taught or chosen - a knowing, a certainty, of the emotions and experiences we share with other

Jane with Jenny and Jim Desmond and Princess
Image Source: Liberia Chimpanzee Rescue & Protection

beings. Like you, one of my most vivid childhood memories involves harm to an insect and the visceral reaction I had. As we shared our childhood stories, I remember thinking oh my gosh, someone who understands how it felt to be a child who knew something as fact - clear and obvious, irrefutable – that frustratingly didn't seem obvious to others.

Our parents, who accepted and nurtured our obsession with animals, despite not quite understanding it all. Our comfort with and certainty about, much to the dismay of many, assigning 'human' emotions and actions to individuals of other species.

And our chimpanzees. Like you, I didn't plan on a life with chimpanzees or have any idea they would become the focus of my world and the symbol for all that I do. I certainly never could have predicted that you and your chimpanzees would take me full circle from my seven year old self whose mother gifted me with your first book to the person I am today.

Jane – thank you. Thank you for sparking the belief that I could have a life like yours. Thank you for trusting in shared sentience. Thank you for spending a lifetime fighting for our cousins. Thank you for your sense of humor. Thank you for your inspiration. Thank you for both the simple moments and the deep realizations. Thank you for the encouragement and support. And, most of all, thank you for being a friend.

I celebrate the day you were born and the gift that is to us all.

With greatest of love,
Jenny

JENNY DESMOND – Founder, Liberia Chimpanzee Rescue & Protection – rescuing chimpanzees in need, keeping wild chimps wild, U.S. Affiliate, Partners in Animal Protection and Conservation

liberiachimpanzeerescue.org

Jane and the Moon Bears

— Jill Robinson —

Whenever I think of Jane I think of the word "indomitable" because it's the description that Jane uses often to describe the passionate determination and spirit of students across the world inspired by her Roots and Shoots campaign. And it's a word that so perfectly describes Jane herself.

It was in 2005 when Jane was in Chengdu, Sichuan Province, as part of a wider "Roots and Shoots" tour of China, and "popped over" to witness a health check on one of our recently rescued group of four bears. An illegal bear bile farm had been closed in a local city by the local authorities, and the bears were now being transferred into our care.

As soon as the truck door opened, we immediately saw evidence of chronic and debilitating mental trauma as the bears rocked back and forth in their cages as if trapped in their own asylum of fear. Their heads showed bleeding, hair loss and scarring from their repetitive face-rubbing against the bars, throughout decades of being caged and extracted of their bile. Their physical injuries too saw Jane's eyes filling with tears as our team carefully offloaded each bear from the truck and carried the heavy cages into the hospital for the first health check they would ever receive in their lives.

We quickly realised that all four bears had received cruel "free-drip" fistula surgeries, that entailed cutting crude holes into their abdomen and gall bladder so that the bile could "freely drip" out and be collected for use in traditional medicine products. Sadly this remains the standard for bile extraction that is accepted by both the industry and the authorities today.

While shocked with their condition, we were able to help the bears after gaining permission from the authorities to offer them professional veterinary

Jane and Mandela, a rescued moon bear in China
Image Source: Jill Robinson

care. We decided that the last bear to be transferred to the hospital would be the first bear to be cut out of his cage because he was so violently stressed and upset. Jane was devastated with his condition. Now sleeping under anesthetic on the makeshift health-check floor outside our small hospital, his thin and wasted form was laid out on the tarpaulin, and told its own horrific story as our vets and nurses worked urgently to assess his wounds and prioritise him for the surgery he would inevitably need to repair or remove his damaged gall bladder. His pads were fissured and scarred from years of standing in a cage, and he had infected wounds and lesions all over his body from lying permanently on metal bars. His beautiful lemony crescent moon of fur only emphasised his bony chest from years of assault and starvation on the farm.

Jane quietly sat on the steps of the hospital, watching our team working, and taking everything in. Just before the health check ended, she did something very special, which moved everyone to tears. Asking for some water to be poured into her hand, she then sprinkled it over the bear's head

After years of exploitation in a tiny cage, Mandela enjoys freedom, fresh air, and the green surroundings of the Animals Asia Moon Bear Rescue Centre in Chengdu, China.

Image Source: Jill Robinson

and christened him "Mandela" in recognition of how, like Nelson Mandela, these bears were so forgiving after years of imprisonment against their will. That moment was something I'll never forget, as she kissed Mandela's nose, then mine, and said a quiet "goodbye and good luck" to her bear.

Like millions the world over, we love and celebrate you Jane for your indomitable spirit and for always giving us and future generations, reason for hope.

Jill Robinson – Founder & CEO Animals Asia Foundation (3 sanctuaries in China and Vietnam seeing peace and contentment for nearly 700 previously farmed bears, and a formal agreement with the Vietnam government to end bear bile farming once and for all).

animalsasia.org

How We Create Change

— Jo-Anne McArthur —

Activism for animals has blossomed and bloomed since Jane Goodall began her journey to help chimpanzees. Efforts by a few caring citizens have expanded over the decades, and we now have a global community, and what's really exciting is that the work is increasingly collaborative and multi-pronged. Our efforts to help animals are now concerted and enduring.

I learned of Dr. Jane when I was a child. Why does she remain a shining light and a hero to me, over three decades later? Because it's even more apparent that her pioneering adventures launched many more dreams and ideas about what we can do for animals. She helped launch mine. Like many young women, I felt empowered by Jane's journey, and I think that millions of us can credit our efforts and successes to her inspiration.

Today, we have the field of animal law. The field of ethology is established. We have journalism that considers and includes animal lives. All animals, not just the charismatic creatures. We have many examples of how animal advocacy has matured, and we can also now say that people are beginning to understand that animal stories are our stories, too. They overlap – are entwined with – the critical problems and solutions of our time.

We've come a long way. I think that's a great thing to celebrate on Jane's 90th birthday.

Jo-Anne McArthur – Photojournalist and Founder of We Animals Media.

weanimalsmedia.org

A rescued pig in pastures full of chamomile at Farm Sanctuary. The very first animal Jane habituated was a pig whom she named Grunter. Jane sat and threw him apple cores, until after two weeks he took these from her hand and she could rub his back.

Image Source: Jo-Anne McArthur, We Animals Media

— Jon Stryker —

Dear Jane,

First, I want to wish you a very merry and happiest of birthdays!

Second, I know you know how important you are to me and have been my entire life, but I will be more specific here: When I was a young boy growing up in Michigan in the 1960's and early 70's, I was completely enchanted with the work you were doing with the chimpanzees at Gombe. As I came to know (and memorize!) all the various names and personalities of those chimps, I began a long and profound journey in understanding the human species and our often-troubled relationship to our planet. Your tireless efforts over the many decades since then continue to influence my life and work. Today, as you know, I am very involved in Great Ape sanctuary and conservation that was very much inspired by you. I even became a vegetarian after your first visit to my home in Kalamazoo way back in the mid 90's. I also have been greatly influenced by your tireless commitment to supporting the human communities adjacent to Gombe and thus have imbedded social justice and economic development into all our conservation efforts at the Arcus Foundation. Your perspective that social justice for people must be integral to how we work to manage the natural world is, I believe, one of your greatest influences and contributions.

Third, the way you show love and compassion to all animals, plants and people is something I have always tried to emulate. Your incredible leadership and advocacy for all living beings has left such an indelible and meaningful imprint on our crazy and beautiful world... and of course on me!

With great love and admiration,

Jon Stryker (he, him)

JON STRYKER – Stream Lone Circle LLC

The Gentle Human Touch that Beacons All to Respect Our Animals

— Josphat Ngonyo —

Animals are more than man perceives them to be – much more than what man has agreed that animals need to be.

This deep silent instinctive realization came to me after I heard a rustling in the bushes one morning while on patrol. I was walking in the bushes in the outskirts of Tsavo East National Park. I investigated to see what it was. My eyes fell upon a struggling baby dik-dik that was caught in a snare. The baby dik dik was struggling to free herself. I bent over and removed the snare from around its neck, it stumbled to its feet and dashed away. But my great urge to free the animal and the actual act itself was not what stood out in my memory of that day. As it sped away from what it perceived to be harm, the dik dik stopped abruptly, looked back and looked at me deeply into my eyes for what seemed like an eternity. It was wordless. It was calm. I held my breath, and my heart skipped a bit. I understood with great conviction what the beautiful baby dik-dik was saying: Thank you. Then just as suddenly, the dik-dik hopped away. That changed my life and vocation forever.

Since that day, I made it a personal mission. I went on to not only work to save animals in distress, but to care for them, protect them and advocate for their freedoms. I embarked on establishing an organization with programs dedicated to rescue animals, international veterinarian programs and place-ment of vets on staff in partnership with Colobus Conservation to treat and heal animals in accidents at the Kenyan Coast. With every rescue, I reflect on sentience and sapience of animals as I perceive the look of deep-seated gratitude in the animals' eyes before they trot to rejoin their herds. This has not only enabled me to understand that animals have different personalities, but it has motivated me as a human being to meet my obligation to protect the animals.

In the noble work, I came to learn of Jane Goodall and have been greatly inspired by her work to protect and advocate for animals and their habitat in such a dedicated and gentle way. In her interaction with animals, I am profoundly awed by her courage, empathy, patience, gentleness, perseverance, determination, tenacity, and independence. I echo her sentiments that given that chimpanzees and many other animals are sentient and sapient, then human beings should treat them with respect. Her people centered approach to conservation has informed my work in my quest to engage and involve women, the youth and children in communities and enable awareness in promoting the welfare of animals. I reiterate her call for humanity to urge for hope, a value that is well expounded on in her books.

Hers was initial studies with the late Louis Leakey in their quest to research and learn about the Great Apes, just as my guardian, the late Rosalie Osborn, did in the past. Mine was an opportunity to volunteer and work with David Sheldrick Wildlife Trust to care for and rehabilitate elephants. We both carry the mantle with great humility. I am beholden to live on this earth with a kindred spirit, as Jane Goodall, in our dedication to protecting the planet's animals.

A blessed happy 90th birthday to you, Jane Goodall! Kila La Heri – Swahili for 'Best Wishes'

JOSPHAT NGONYO – Executive Director, Africa Network for Animal Welfare

anaw.org

Tribute to Jane Goodall

– Joyce D'Silva –

Like so many others, I first knew of Jane as this brave woman who had lived with chimpanzees in Africa and recorded their lives, their actions and their emotions. When I heard her speak about how academia had insulted her research methods, I felt such admiration for her courage in pursuing her course regardless.

At Compassion in World Farming's conference on animal sentience in 2005, Jane told her story. As she said then: "Humans are unique, chimpanzees are unique, elephants are unique and pigs are unique".

Jane had already been deeply influenced by Peter Singer's book, *Animal Liberation*, published in 1975. "The next time I saw meat on my plate I thought 'ah, that symbolizes fear, pain, death.' And I didn't eat any more ever again." (vegofwa.org/2023/02/09/jane-goodall-publishes-her-own-vegan-cookbook/)

Nowadays, Jane has moved even further, publishing a vegan cookbook, "Eat Meatless" in 2021 and addressing the huge environmental problems caused by factory farming.

Although she lived so closely with the chimpanzees, Jane is well aware of the dangers of disease transmission – both ways – from animal to human or vice versa. At a Compassion in World Farming event in 2020 she said: "Our disrespect for wild animals and our disrespect for farmed animals has created this situation where disease can spill over to infect human beings. We have come to a turning point in our relationship with the natural world."

The solution, she says, lies with us: we should "eat less meat, or no meat, and turn to plant-based diets". (weforum.org/agenda/2020/06/jane-goodall-coronavirus-humanity-natural-animals-covid-finished)

The Jane Goodall Institute does wonderful work all over the globe, mirrored by her youth groups, Roots & Shoots, where young people work to improve the lives of people or animals or help the environment. Jane constantly travels to visit these groups and encourage them in their good work.

This is community-based work, but Jane has also become an outspoken advocate for change. When Compassion asked her to support our campaign to stop the EU funding promotion of meat etc., she wrote: "The European Commission's recent food and cancer policies show it well understands the need for a shift away from animal products towards more plant-rich diets, but its policy for funding food ads doesn't yet reflect this: it seems important that these conflicting messages be brought into line. We are calling on the Commission to reform its EU farm products promotion policy so that it provides support and incentives for the crucial shift to more plant-based diets in Europe. This will benefit people, animals and the planet."

I spoke on the phone to Jane during COVID, when she was unable to leave her home. Apart from the permitted 1-hour walk, she was at her computer working "usually until 2 or 3 in the morning".

At Compassion's Extinction or Regeneration conference in London in May 2023, Jane emphasised that "we are just part of the natural world". She admitted that investigative films of animals in factory farms "have kept me awake at night". She knows that each animal is an individual who "can feel depression, fear and pain".

Compassion in World Farming is honoured to have Jane as a Patron, and she echoes our call for gentle, regenerative farming, an end to factory farming and a move towards plant-based eating.

She will always be one of my heroes!

JOYCE D'SILVA – Ambassador Emeritus, Compassion in World Farming. Author of *Animal Welfare in World Religion – Teaching and Practice*

ciwf.org.uk

Shared Values and a Long Friendship

— Keely and Pierce Brosnan —

It is an honor to have had the pleasure of knowing Dr. Jane Goodall, DBE for over a quarter of a century. Jane's unrelenting devotion to safeguarding the natural world has left an indelible mark on both the animal kingdom and human hearts around the planet. There is a beauty in her humanity and humility as a messenger of peace. She is both elegant and passionate as well as a force of nature in her own right. Jane's ability to bridge the gap between scientific knowledge and public awareness has ignited global conservation efforts, inspiring individuals to deepen their connection with the animal realm and nature.

Early in her career, Jane recognized the interconnectedness of all life forms and the need for conservation. Jane's tireless efforts to halt the illegal wildlife trade have played an instrumental role in raising awareness of the devastating consequences of poaching and habitat loss. The Jane Goodall Institute, which partners with local communities to establish sustainable livelihoods and holistic approaches to conservation, stands as a model of excellence.

Among many memorable moments with Jane, in 2005, we had the privilege of hosting an afternoon tea in support of Roots & Shoots – a youth-based educational program which fosters environmental awareness among children and teens. It was unforgettable listening to Jane bring to life her captivating stories of Gombe for our local community in Malibu.

Subsequently, in 2008, we hosted the Roots & Shoots' Day of Peace in Los Angeles, a celebratory gathering which nurtured a sense of responsibility and empathy for the earth, animals, and humanity.

Jane's journey as an animal advocate has been defined by groundbreaking research, unwavering dedication, and a longstanding commitment to conservation, reshaping our perception of chimpanzees and other primates by highlighting their intelligence and emotions as well as our similarities and interconnectedness. Through advocacy, education and outreach, Jane has inspired a movement that transcends global boundaries, fostering a collective responsibility to protect our planet and all its inhabitants. It is always a joy to be in Jane's company, to learn more about her remarkable encounters and achievements and to stand in awe of her resolve to make the world a better place for ourselves and our children.

Jane's indomitable spirit, hope, and legacy will continue to inspire generations to come – including our sons Dylan and Paris – to remember that our future is inextricably linked to the well-being of all living beings.

Her words of wisdom ring true everyday: "What you do makes a difference and you have to decide what kind of difference you want to make?"

KEELY SHAYE BROSNAN – environmentalist and film producer

PIERCE BROSNAN – actor

Jane in Taiwan

– Kelly Kok –

Whenever Jane visits Taiwan, a special kind of fever descends upon us. Our team, like any other team, works hard to stock up special drinks, chocolates, nuts, carrots, cabbages, a bit of curry and rice, and of course plenty of fruit. Then the glorious days start, as we work alongside her, and watch her speak with wisdom and humour, and with kindness, in every occasion, and to everyone.

We always put Jane in the same hotel for her visits, always in room number 1934, the year of her birth. The suite is a home away from home for Jane, and it is filled with many happy moments, memories and nightcaps with beloved friends and supporters of JGI.

After her lectures, she would stay for hours to autograph books, T-shirts, caps, binoculars, and even the arms of her ardent fans. We are not allowed to turn anyone away. One of our tasks is always to plot the perfect escape route for her to leave a speaking venue. But of course, determined fans will always find Jane!

When a disastrous earthquake hit Taiwan in September 1999, I asked Jane if she could postpone her visit to us that year, as we were unable to fundraise for the institute during that period of need. She simply said, "How could I not come to my friends in need?" She came, and together we visited temporary shelters in the earthquake stricken areas, and hosted Roots & Shoots events that helped the afflicted communities. Working alongside Jane, we saw how she listened, how she responded with wisdom and compassion, and how her empathy worked wonders.

Journalists baited her with questions, "Do you think our government is doing enough? Do you think more should have been done?" And she answered gently, "When bad things happen, the first thing we do is to try to find someone to blame, so that we can direct our pain at something else. It is precisely at this moment that we need to stand together and help each other." The entire room of journalists was quiet for a long time.

We love it when Jane joins our Roots & Shoots Animal Parades. It is full fledged party time, with everyone dressed up as animals to celebrate the diversity of the natural world. Well-meaning artists once created a carriage for her, so as to spare her the hassle of walking. Of course Jane said no! While the animals walk and prance she will walk alongside them, meet everyone, and enjoy the fun.

We carry this image of her with the animals, laughing, dancing, and carefree, in our hearts. We are blessed to be part of your team and dream.

Happy 90th birthday dearest Jane! Together we can, together we must, and together we will!

With all my heart and love,
Kelly

KELLY KOK – Executive Director of the Jane Goodall Institute Taiwan

— Leonardo DiCaprio —

Jane. Happy 90th birthday. Too few people are granted the privilege of getting to look their heroes in the eye and stand with them as I have been humbled to stand with you. I am grateful to be able to call you not just a hero but a true friend.

Your name alone invokes the power of one person's voice for animals. You have made an indelible impact on our planet, and we are grateful for your tireless efforts in being a fierce protector of every living creature. Your ability to always choose grace and compassion serves as an example for the world to follow, something I think about on the days when hope for the future of our shared Earth feels hard to reach for.

David Greybeard,
the first among the
Gombe chimpanzees
to accept Jane's presence
in the early days of her research.

Image Source: Hugo van Lawick

From anyone else the message of hope, and the insistence that we must still hold onto it, might feel like an empty platitude. From you, it is a reminder that pessimism is the easy choice, the white flag of surrender raised up in the face of challenges that can seem insurmountable. Hope is more difficult, but our planet is worth it. This is a lesson I have learned from you, and one that I look to as a North Star for guiding me in my own work driving environmental action with frontline partners around the world, and as a fellow advocate for the species whose future rests in our hands. You have given a voice to animals, which has inspired millions to have genuine compassion and understanding of our fellow species who we can help protect.

Jane, you have inspired me through your words and your work years before we met and continue to do so every day that I am fortunate to call you a friend.

You held a mirror up to humankind and showed us the truth of ourselves through the faces and the behavior of our closest cousins, the Chimpanzees. With the help of David Greybeard, you changed forever how we understand our place in the world, and for that we are grateful.

LEONARDO DICAPRIO – Actor and environmentalist

Jane Goodall: A Legacy of Compassion and Connection

— Lesley Day —

On the momentous occasion of Dr. Jane's 90th birthday. I find myself reflecting on the incredible impact this extraordinary woman has had on our world. I feel both honored and overwhelmed to have the opportunity to pay tribute to her once again.

Dr. Jane has graced our sanctuary, now named Freedom for Great Apes, not once but twice. I vividly recall her first visit in 2005, when she arrived in Bend, Oregon, and was greeted at the airport by a sea of banners and flags held high by her devoted Roots and Shoots group from Sisters, Oregon. The enthusiasm and admiration were palpable as she interacted with the young members, who displayed their projects later at the high school where Dr. Jane was talking. Dr. Jane's unwavering commitment to connecting with every possible person time will allow, especially the younger generation, is a testament to her selflessness and dedication.

During her tour of our sanctuary, I watched with anticipation as Dr. Jane approached the enclosure where our adult male chimp, Herbie, was waiting at a barred window. I was a little concerned for her safety as she approached the window. I totally underestimated the profound understanding that exists between our closest relatives and this remarkable woman. With outstretched hand and unwavering composure, Dr. Jane drew closer to Herbie, and in a moment that will forever be etched in my memory, he reached out with one gentle finger to touch hers. It was a profound testament to the trust and connection she effortlessly establishes with these magnificent creatures.

What truly sets Dr. Jane apart is her unwavering passion and unyielding mission to protect and conserve our natural world. Her calm and serene demeanor belies a steely determination to combat the environmental challenges that threaten our planet and its inhabitants. From her groundbreaking research on chimpanzee behavior to her tireless advocacy for conservation and animal welfare, Dr. Jane has inspired countless individuals, including me, to take action and make a difference. Her impact is immeasurable, and her influence will continue to reverberate throughout future generations.

As we celebrate Dr. Jane's 90th birthday, we are reminded that she is not a mere mortal but an indomitable force that will forever be intertwined with the very fabric of our world. Through her extraordinary life's work, Dr. Jane has demonstrated that one person can indeed change the world, and her influence will continue to inspire and empower us all.

Happy 90th birthday, dear Jane. Your light will continue to shine brightly and guide us on the path towards a better and more harmonious world.

LESLEY DAY – Founder of Freedom for Great Apes in 1995

freedomforgreatapes.org

Connecting Our Brains and Hearts
with Satellite Maps

— Lilian Pintea —

D ear Jane,
In December of 2000 we met in your house in Gombe, in a dark room lit by candles, sipping whiskey, and looking over a new source of data – images of Gombe observed from space by NASA satellites, and even images before satellites were available, acquired from planes in 1947 and 1958, just few years before you arrived in Gombe. We were transported in time by the black and white photos, seeing a landscape that only remained in memories and old photos, a landscape where Gombe was still connected to large forests and woodlands along the Great Western Rift Valley in Africa. I remember you saying, "This is magic" and I felt that we are at the beginning of a special journey.

Growing up in Moldova in the former Soviet Union, your translated books on Gombe chimpanzees captured my childhood imagination and inspired me to dream to become a scientist and study wildlife in Africa. Behind the Iron Curtain it seemed like an impossible dream. A few years later, as a graduate student in the USA, I was searching for a PhD project in wildlife conservation that I could support with my growing technical skills in applying innovative remote sensing and GIS technologies. In return I was eager to learn how science, data and knowledge could support real-world decision making and conservation in practice. I was fortunate to have few choices such as studying tigers in Thailand, monitoring forests in Brazil or chimpanzees in Gombe National Park, Tanzania. It seemed like destiny.

As a PhD student in Conservation Biology at the University of Minnesota working under the direction of your former student Dr. Anne Pusey, I was fortunate to connect your long-term chimpanzee data with satellite imagery to understand how chimpanzee behaviour, habitats and human land uses changed between 1972 and 2003 inside and outside the park. However, the most important things that I learned came from using this data and geospatial technologies with the local communities to support reforestation as part of TACARE project that you, George Strunden, Emmanuel Mtiti, and others initiated in 1994.

My favorite story is when we first brought 1-meter, high-resolution Ikonos satellite imagery to the villages in 2003. We were kneeling on the ground around large paper prints of satellite maps when one woman pointed to her house, the path that she was taking to her farm and even the large tree, where in its shadows, she was leaving her baby to sleep while she was farming. She then showed us where she stops to collect her firewood and the path that she takes to get back home. As she was sharing her story, all of us around that map also mentally travelled with her through the imagery, not only understanding how she valued and used natural resources but also connecting our minds with our hearts by experiencing a day in her life with compassion. I was struck by her story and remember it to this day because it was the moment that I realized that thanks to TACARE, these maps on the ground created a common language and understanding between us.

Today, similar high-resolution satellite imagery help us see that, thanks to TACARE, many of the forests and woodlands outside Gombe are in the process of being restored through natural regeneration as part of Village Land Forest Reserves set up by local communities and informed by science and data. Thank you, Jane, for sharing your wisdom that "Only when our clever brain and our human heart work together in harmony can we achieve our true potential".

LILIAN PINTEA – Vice President Conservation Science, the Jane Goodall Institute

Jane-Magic!!!

— Lorenz Knauer —

Memories of filming *Jane's Journey**
January 2009: We had no more than four precious days with Jane in Gombe. And after three days, still not a single shot of chimps!! Just a few baboons... My heart began to sink. On our last morning my heart sank even lower as the park rangers told us the chimps were way up in the hills, no chance for us to film. To my puzzlement, Jane just smiled calmly. All of a sudden, excited crackles from a walkie talkie, signaling that a large family had moved quite close, up towards the Peak. I never ran up a hill so fast in my life carrying heavy camera-equipment... And within the next three hours we were gifted with the most gorgeous chimp-Jane-footage we could have imagined! Jane quietly said: "You know, the IMAX-team that was here for months on end didn't get half of what you got this morning!" We titled this experience: "Jane-Magic".

The Waterfall in Gombe

Jane said: "Lorenz, you are crazy! A ten-meter camera-crane that has to be carried up all the way to the waterfall" Well, we actually got it up there, set it all up, were ready to go and then: It began to rain. And rain. And rain. My heart sank again... time was running out, we only had that afternoon to get this shot! Jane very calm and smiling: "At 5 p.m. the sun will be perfect, shining right on to the waterfall." That was at 3. The rain just kept pouring.

---

* The documentary *Jane's Journey* (2011) is available on Amazon (Prime) and in iTunes.

A chimpanzee family in Gombe

Image Source: André Zacher

Jane among some of the Gombe chimpanzees

Image Source: André Zacher

At 4.15 it slowed. At 4.30 it morphed into a slight drizzle. At 4.45 the clouds parted and we began to rehearse the complicated shot with Jane walking towards the waterfall... At exactly 5 p.m. the sun shone right onto the cascading water... it was truly magical, even more beautiful than if it hadn't rained... the leaves were wet and glistened in the sunlight... we canned one of the most powerful shots of the entire film. And Jane just smiled her Jane-Magic smile.

The Hippo-Pool

One of the most eerily beautiful places I have ever been to – deep in the south of Tanzania, close to the border with Mozambique. Many, many hippos, huge crocodiles, exotic birds of all kinds... and the pool itself bathed in absolutely unreal, green light... We were rehearsing a long Steadicam-shot with Jane and Yahaya, the Hippo-Whisperer...everything seemed peaceful, the hippos basking lazily in their pool when all of a sudden – a horrible, crashing sound...My heart not only sank but practically stopped beating when I realized what was evolving within a matter of seconds: two huge bulls had lumbered out of the pool and were now charging in our direction – exactly in their path stood Jane, right next to Yahaya. My brain froze, this is it: "Jane is going to be run over by two charging hippos and it's all my fault!!!" The hippos stormed by, just inches away from Jane and Yahaya and disappeared back into the pool with great crashing and splashing. With her Jane-Magic smile she asked me: "Did you know that more people in Africa die from hippos than from crocodiles?" Unforgettably Jane.

Lorenz Knauer – Filmmaker and President of the Jane Goodall Institute Germany

lorenzknauer.com

The Power of Names

– Lori Gruen –

There is no one more iconic than Jane Goodall as a force for compassion for our fellow creatures. As a tireless advocate not only for chimpanzees, but for all animals and their habitats, she has made the wonders of the natural world more vivid to all of us. As a young girl Jane was my hero, her courage and determination inspired me and through her boldness, she showed me that it was OK to do things one wasn't expected to do and to name and care about animals not simply as numbered "objects" to be studied. I'm an older woman now, and Jane is still my hero. When I developed two websites honoring chimpanzees, *The First 100*, that names and describes the chimpanzees that were first used in research in the US and *The Last 1000*, that tracks the final chimpanzees used in research as they make their journey to sanctuary, Jane's audacious decision to name David Greybeard, Flo, Fifi, Figan, Gremlin and so many others was always on my mind.

As I dug through archives, wrote freedom of information requests and other letters of inquiry in my work to discover the names of these chimpanzees, most of whom had names due to Jane's innovation, the power of naming others became even more apparent to me. The first 100 chimps used in research are all now deceased, but in bringing their names and their relationships with other chimpanzees to light, who they were as individuals – as mothers and fathers and siblings, with their own likes and interests and idiosyncrasies – could be memorialized. In listing the names of all 1000 chimpanzees who, in 2013, were still in laboratories and turning each name green as they made it to sanctuary I tried to honor each and every one of them as individuals.

Jane's efforts to help end invasive chimpanzee research and allow the last

1000 to move to sanctuary is yet another reason to honor her too. We are so lucky to have a person like Jane among us. My birthday wish for Jane on her 90th birthday is that she could live forever and continue to inspire more and more youngsters the way I continue to be inspired by her powerful vision of compassionate co-existence with other beings on our increasingly imperiled planet.

LORI GRUEN – Wesleyan University; Author of *Ethics and Animals: An Introduction* and *Entangled Empathy*

lorigruen.com

first100chimps.wesleyan.edu

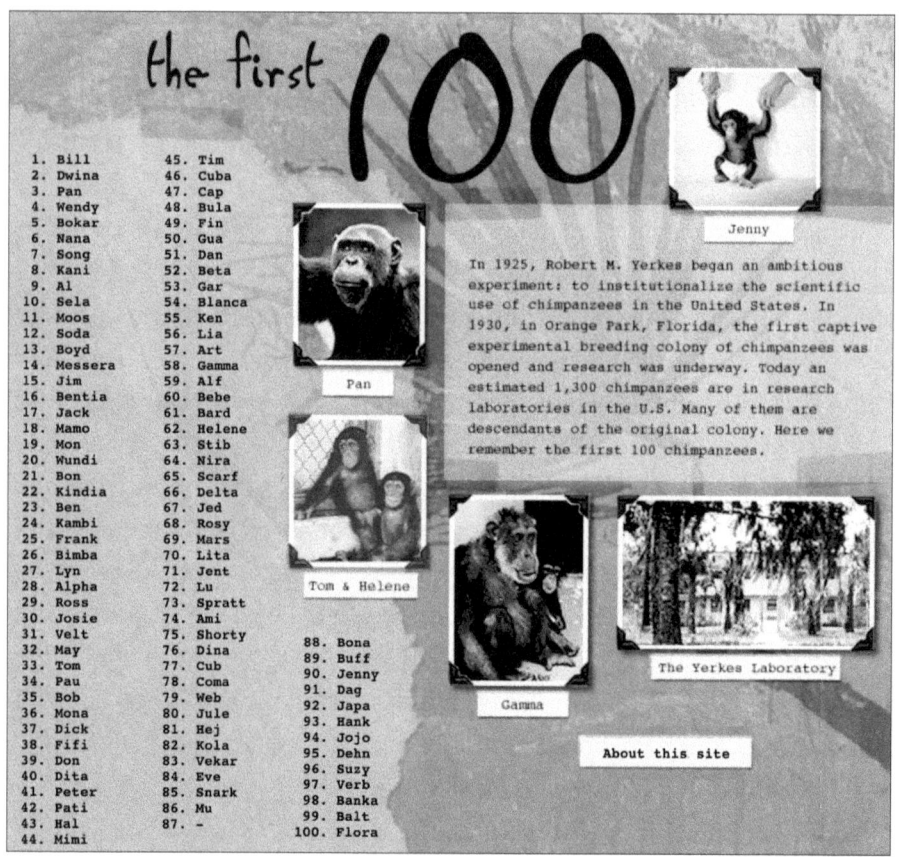

Remembering the first 100 chimpanzees used for research in the USA

Image Source: Lori Gruen

A Note from Marc Benioff

– Marc Benioff –

I'll never forget the first time I met the legendary Jane Goodall. It was years ago, and we were at a conference focused on the plight of our oceans. Jane asked me what I was working on, and she listened kindly as I spoke about cleaning up plastic from the oceans and protecting the coral reefs.

After I was done speaking, Jane looked me in the eyes and said, "Marc, it's nice what you're doing for the oceans. But what are you doing for the forests?"

I didn't have an answer.

At that time, I didn't fully understand the connection between the oceans and the trees, between blue and green carbon – the interdependence of the total ecosystem. Jane challenged me to help protect the carbon-sequestering land forests that absorb billions of metric tons of carbon and harbor 80 percent of the world's terrestrial biodiversity. Our conversation began a journey that led to 1T.org, a program to help re-green our planet and conserve, restore and grow one trillion trees by 2030.

Jane is one of my true heroes, and I'm not alone. Around the world, generations of advocates and activists – especially young people – have been inspired and mentored by Jane, by her fearless advocacy, and, above all, by her unyielding embrace of hope itself.

Through her more than six decades of extraordinary work and her personal example, Jane reminds us all that we belong to a greater family – we have a "spiritual connection" with nature – and we all have a responsibility to improve the state of our world. As Jane says, "Together we can bring change to the world, gradually replacing fear and hatred with compassion and love. Love for all living beings."

Whenever I am with Jane, I feel I am with one of the wisest people on the planet. When I look into her eyes, I see the future of our humanity in harmony with the animals around us. I see our spiritual connection with all of nature. I see hope.

Happy birthday, Jane!

MARC BENIOFF – CEO Salesforce

Jane Goodall: Messenger of Peace and Hope and Master Storyteller

— Marlon H. Reis —

I first met Dr. Jane Goodall in March of 2023, when she came to speak in Denver, Colorado. My dear friend Dr. Marc Bekoff told me the good news and asked if my husband, two children, and I would like to attend. Miracle worker that he is, Marc arranged for us to meet Jane for a few, precious moments before she went onstage.

It's no exaggeration to say that when I came into my role as First Gentleman of Colorado, and I found myself suddenly possessed of a platform and an audience to hear messages of my choosing, I thought of Dr. Goodall and how she'd helped so many people find reason to care about each other, and the world we all share.

As an ambassador for the natural world and the countless precious, non-human lives with whom we share our planet, Jane has inspired and taught me that to live a life of meaning, one must do so purposefully and with heartfelt conviction. If all I had were second-hand accounts to form an impression of what Dr. Goodall must think of humanity and its often-unwelcome advances in the natural world, I might easily assume her to be a cynic.

But one doesn't assume such things about Jane Goodall. With the eloquence of one who has spent a lifetime moving quietly through sacred places, she tells us that if we want people to stop cutting down the Amazon, then we must offer the promise of livelihood in another form. He who swings the axe does so not out of spite for the forest, but because it puts food on the table for his family. So, it follows, it is not for the people of Brazil to solve for the world's misplaced values, but for those of us living lives of plenty, who likely never bother to imagine that we are the true commissioners of tropical deforestation – the unreflecting consumers who demand without understanding the true extent of what we ask of the Earth.

Dr. Goodall has a rare talent for making the world small; for bringing what is distant and foreign, near and personal; for reminding us that, however powerless we may feel, the way we live our lives as individuals does indeed matter in the grand scheme of things. Every life, great or small, is part of that grandeur.

Jane the scientist is Jane the storyteller. In a Jane Goodall story, there are no "supporting roles". Every being, every place, is a fully realized character. And one cannot tell the story of one without telling the story of the other, because all life is interconnected. And it is only by understanding our fellow nonhumans that we can understand ourselves.

For the great many of us who know little of science, Jane gives us reason to care in telling the tales of animals she's named and known; who some days welcomed her and some days did not; whose lives had beginnings, middles, and ends; all of whom wrestled with the self-same proposition of hanging on in a world of hope and fear, joy and sadness, and the intensity of uncertainty. The story of life is the story of being here, human and nonhuman alike.

MARLON H. REIS – First Gentleman of Colorado

Jane Has Always Been Important to Me

— Mary Ford —

I don't remember a time when I didn't know about Jane Goodall. It feels like she has always been a part of my life, always been an inspiration. I do remember the moment when I had a clear realization of my own deep love of animals and how that connected me to Jane. I was nine years old and had been very sick for many weeks, but I was finally well enough to sit on the porch and hold our cat in my lap. I thought, "I will be okay as long as I can be with animals – just like Jane Goodall!" I felt that I was in very fine company.

Knowing Jane's story also gave me the idea that I could study animals and helped me find the courage to explore that kind of career. Upon graduation from college, I got a job with Dr. Cheryl Knott, supporting her study of wild orangutans in Indonesia. It was fascinating to follow the orangutans, recording their behaviors and diet, using techniques pioneered by Jane. But while I was there, I realized that I really missed working with children. I have always loved children and had volunteered in several nature centers and classrooms. As I was marveling at the wonders of the Bornean rainforest, I kept thinking about how important it is for young people to experience nature's magic and be involved in protecting it. I was thrilled to learn that Jane also immensely values the voices of children and that she had created the Roots & Shoots program to give them a way to make a difference. Roots & Shoots immediately became my north star.

Ever since my return from Borneo 25 years ago, I have worked with children, finding ways for them to connect with nature and act as changemakers. I am so grateful that my career eventually led me to the Jane Goodall Institute USA, where I support Roots & Shoots and Jane's work across the

Jane with Roots & Shoots kids in Nebraska, 2005.
Image Source: Thomas D. Mangelsen

Jane connecting with a young girl in Washington DC, 2008.
Image Source: Thomas D. Mangelsen

country. Roots & Shoots members take action through local projects that benefit the environment and improve the lives of people and other animals. Importantly, the projects are led by the young people themselves, who explore their communities to determine needs and decide how they can help. This results in insightful and innovative projects, ranging from planting mangroves along the coast of Florida to making dog toys for pet fosters to collecting board games for family shelters.

Jane has been important to me since I was that little girl sitting on the porch. It is a tremendous honor that I now work with today's young people, bringing Jane's message of hope in action to a new generation. As Jane says at all our Roots & Shoots gatherings: "Together we can! Together we will! Together we must!"

MARY FORD – Senior Director of Roots & Shoots USA

Conoco to Chimpanzees

— Mary Lewis —

❝ Come to lunch with Jane Goodall" urged Dino Nicandros, then President of Conoco Inc and so I found myself across the table from Jane in Phoenix, Arizona. In 1990 there was no Google – I knew little of Jane's work apart from a small display featuring the sanctuary that Conoco was creating for 25 orphan chimpanzees in the Republic of Congo.

Jane's first question that day was "what is Conoco doing about environmental education?" There was a long silence... and to fill the vacuum I talked about 'Understanding our Environment' a programme we at Conoco were creating for European schools. That day changed my life and 33 years later I'm still attempting to keep pace with Jane. Jane's life has entailed countless cities, countries and venues from cathedrals to sports arenas, theatres to crowded small rooms, parks and fields and the tours have always included creatures. Whether it was a campaign for prairie dogs, the plight of cougars etc. or dogs encountered or loaned, worms rescued from sidewalks, conversational parrots, painting pigs, horses, chickens, frogs, meerkats and of course so many primates and chimpanzees. Then there was the Bournemouth Robin who daily sang to Jane under her Bournemouth birch tree each lunchtime as she sang to him... and so many birds lured into hotel rooms with trails of crumbs over the years.

'Every individual makes a difference' and what a positive difference Jane has made to so many many lives human and non-human. In the early days Roots & Shoots was described as 'nebulous' and Jane was told that her plan for an holistic programme with "projects to benefit people, animals and the environment" would **never** catch on. More than 30 years later young people

Jane and Mary Lewis

Image Source: Thomas D. Mangelsen

of all ages in more than 65 countries are inspired to do something to make this a better world. The passion has been passed on to successive generations!

Jane – thank you for the amazing programmes you have created, the countless talented colleagues you have inspired in the Jane Goodall Institutes around the world and for the inspiration you have been for the thousands of people whose lives have been altered by Jane magic. Whether attending a lecture, receiving a call or letter, through a chance meeting in an airport or on a train, your passion, humour, patience, persistence and inspiration has reached and benefitted human and non-human animals in countless ways. YOU show us the world we COULD have and inspire us to create the world we SHOULD have. Thank you dear Jane!

MARY LEWIS – Vice President, Outreach, Office of the Founder Global, Assistant to Dr. Jane Goodall, the Jane Goodall Institute USA

Good on Her Word

— Mary Peng —

Jane Goodall is renowned for her ground-breaking research on chimpanzees and wildlife conservation to protect these magnificent primates. But through my conversations with individuals and working with Roots & Shoots groups, many asked me what about the more familiar companion animals in our homes such as dogs and cats? We may now have the opportunity to see wildlife on TV and the internet and occasionally in zoos or sanctuaries, but the opportunity for up-close observation is limited for most of us. Our first encounter with animals is usually through the pets we have in our homes. I asked Jane during one of her visits to China how we could bring the companion animals into our messages and remind the community that JGI's mission is not only limited to wildlife but all the animals with whom we share the space.

Riding in the car with Jane in Beijing on our way to her next speaking engagement, Jane shared with me her childhood experience of how her dog Rusty, was her introduction to loving animals and the profound impact he had on her throughout her life. Jane reiterated that the mission of JGI was not just limited to chimpanzees and wildlife but to all animals.

After we arrived at the venue and Jane was introduced, she delighted the audience with a chimpanzee pant hoot, which she also taught the audience. Jane opened her remarks with the story of her dog Rusty and how her deep love for her childhood dog inspired her love for other animals and the chimpanzees she would later work with in Tanzania.

As Jane spoke about Rusty, my heart was filled to bursting. I was so happy and so deeply touched. Jane heard me and showed me my concerns were

Jane playing with Tom's dog, Loup, in 2002.

Image Source: Thomas D. Mangelsen

Jane and her childhood friend, Rusty, who taught her that animals have personalities, minds, and feelings.

Image Source: Courtesy of the Goodall Family

valid and important. She took immediate action to address those concerns. Over time, I have watched Jane in her interviews and videos bring up her love for Rusty and companion animals and how all the animals mattered. I try to follow Jane's example, and each time I speak with pet owners and stray pet rescuers, I also encourage them to remember the wildlife animals in our midst and the impact we can have on their lives through our words and actions.

I love Jane Goodall. She is a friend, a leader, a mentor and a humanitarian who brings out the best in all of us. Happy Birthday Jane!

MARY PENG – CEO & Founder of the International Center for Veterinary Services (ICVS). Director, Jane Goodall Institute China. Trustee, Jane Goodall Institute Global

Note From the Editors: ICVS was the first full service, international standard animal hospital registered as a wholly owned foreign enterprise (foreign enterprise registration) in China in 2006. As such ICVS helped to raise the standards of veterinary medicine for companion animals in China and continues to do so every day.

Our North Star

– Melody Horrill –

A wild, solitary, injured dolphin helped me heal from my emotional wounds, but it's Jane who gives me and so many others hope for the future of our planet and the precious species we share it with.

I first knew of Jane as student at university. My lecturer was studying wild dolphins and Jane was his role model. He, like her, documented the mammals behaviour by observation. Through volunteering with him, I met a lonely, disfigured dolphin called Jock who taught me trust, unconditional love, acceptance, and forgiveness – lessons I craved after a childhood wracked by domestic violence. I soon became a vocal advocate for them and their ocean home.

This passion led me to become an environmental journalist. That was when I first met Jane in person, at a fund-raising function for the local zoo. I was struck by her gentle, caring spirit and firm resolve. She radiated warmth, compassion, and a quiet determination to help others connect with, and conserve the natural world. She reaffirmed to me that we are intermeshed with other species – plants and animals – as they are with us. We are all a part of the living, breathing matrix of life which is dependent for its survival on all its parts, not just one. That short, authentic chat had a life-long impact on me.

Years later when I wrote my memoir *The Dolphin who Saved Me* I sent it to Jane, hoping she might endorse it. She did! When I read her words I cried – they were so beautiful, gracious, and heartfelt. A few days later, I received a personal email from Jane asking if I'd be interested in joining a new Committee, formed from her deep concern for the welfare of captive dolphins and whales and desire to explore safe ways to return them to their true home

Jock, the dolphin who helped Melody Horrill overcome her childhood trauma. Jock was later successfully integrated in a wild pod of dolphins. Jane asked Mel and Koen to set up JGI's Cetacean Committee, which aims phasing out the keeping of dolphins and other cetaceans in captivity and supports seaside sanctuaries.

Image Source: Martin Jacka

– the ocean. I was deeply honoured and accepted immediately. The issue is so important and resonated so closely with my own heart. Although not a scientist, I hoped I could add value.

That's the thing about Jane, she sees the potential in people from all walks of life, with different experiences and backgrounds and connects them. Titles don't matter. It's all about bonding and uniting people towards a common goal – caring for our planet. Since becoming a member of the JGI Cetacean Committee, I feel as I have been welcomed into the loving arms of a special family – Jane's family.

Thank you Jane. You exude love and hope which illuminates the darkest corners of our world. It's the invisible glue which binds us, helps us find strength and courage to strive to create a better place for all living things. Long may you shine. The world needs you.

MELODY HORRILL – Author, journalist & ocean, earth, animal lover. She is the author of *The Dolphin Who Saved Me* (Greystone Books, 2023). She is also Director of the Dolphin Research Institute (Australia) and Co-chair of the Jane Goodall Institute's Cetacean Committee.

melodyhorrill.com

— Merlin van Lawick —

To my beloved grandmother Dr. Jane Goodall,
Ninety years ago today, the world was graced with a remarkable soul
– yours. Your universal message resonates deeply: to treat others as we wish
to be treated. But it's your message that calls to extend that compassion and
respect to every life form.

When we pause to observe genuinely, it's breathtaking to see how intricately every species is woven into the tapestry of life. Every organism, no
matter how minute, plays a vital role in the collective consciousness of our
planet. While we are each unique, we are undeniably intertwined in this
wondrous journey of existence.

Yet, it saddens me to acknowledge that our actions are, more often than
not, harming the very world we call home. How can we, the most advanced
beings on Earth, threaten our only home? The complexities behind this que-
ry are vast, influenced by deep-seated societal norms, economic drivers, and
political systems that have shaped our worldviews for ages. Yet, at the heart
of these vast constructs are individuals— each with emotions, hopes, and
dreams. This is where true change must begin.

No matter how daunting the question, we must search for answers – for
our survival and that of every creature on Earth hinges on our choices and
actions.

I vow to you that I will use this clever brain of mine to find solutions
for us to evolve in harmony with the natural world and this compassionate
human heart of mine to seek the like that connects all life forms. I'll strive to
find harmony between humanity and nature.

You've instilled in me a profound truth that I have the power to shape the world every day. While the magnitude of global challenges can be overwhelming, you've taught me the value of small, immediate actions. It's about the choices I make daily, and the compassion I show to every living being, regardless of form or function. Every small act sparks hope, and together, we can create monumental change. I urge the other readers to consider that small actions collectively make significant differences. There are so many other people out there who care as much as you do for the future of our planet.

And finally dear beloved, as I pen this message, I am flooded with memories of our time together - your stories, our shared laughter, and the countless lessons you've imparted. As another year adds to your incredible journey, I want you to know that your legacy isn't just in the great work you've done, but also in the hearts you've touched, like mine. Here's to you, to the adventures we've had, and to many more memories together. Happy 90th birthday! With every beat of my heart, I thank the stars for a grandmother like you!!

Merlin van Lawick – Jane's grandson

Happy Birthday to Jane the Siren

— Olympia Ammon—

Turning 90, what a blessing, and a what a nice opportunity to thank you for who you are !

Wow! What a privilege it is to be alongside 89 other top secret sentient beings who have been selected to submit a gift of appreciative words for your 90th birthday! Jane, you are a master at storytelling, needing very little prep for anything, and your words are always met with rapt attention, mesmerizing your audience instantly, like one of the sirens in the Odyssey. So how can I possibly write something that is as good as something you would say? The answer is I can't! So instead, for your birthday, as a fellow Aries, I am going to tell you what I think makes you so magical.

Jane, there is nothing I love more than your laugh. I usually do this by speaking so fast that you have no idea what I am saying (this is very often). Other times it's by asking what birds are outside of your window or how we are both so mortified at a recent current event or bad headline. And the reason I love to do this is that when you are laughing, that is when the energy transfer happens between the two of us, and we experience common humanity. And this happens with everyone you connect with. Your best stories ALWAYS include laughing by both parties. Have you thought about that? Laughter brings us closer than so many other experiences.

I remember once you recorded a video of yourself giggling for some purpose or other (was it National Let's Laugh Day? I'm cracking up just trying to remember this!), and that is my favorite video you have ever made. Honestly you could have asked me to do anything after watching that video and I would have done it (except voting for Trump).

Jane, I am convinced that the contributions you have given to science

Jane and Olympia Ammon in Paris

Image Source: Olympia Ammon

and our planet have been absorbed by millions of people because of your underlying love of joy. It's the medium for your message. You create a space where laughter is cultivated and joy can safely enter. I wonder sometimes if you once laughed with the chimps or shared a moment with them when they, too, felt a spark of joy. Maybe it was then that they playfully revealed their tools to you...

I have determined that laughter is why I am lucky enough to be one of your candles in this book. It's how you and I best communicate. Siren Jane, thank you for showing me and everyone else who has met you how to welcome laughter into every conversation, nurture it, and then magically transmit a message through it so that each of us on this suffering planet know that we make a difference. I sooo love you!

Here we are pictured together above in Paris in November 2022 at a Jane Goodall Legacy Foundation board meeting. It is still my one and only selfie with Jane after years of working together, and I covet it, just like she does Mr. H! We were both laughing when I took it, naturally!

OLYMPIA AMMON – Executive Director, Jane Goodall Legacy Foundation

janeslegacy.org

facebook.com/watch/?v=475293943748442

Sent on a Mission

– Patrick Flynn –

Several years ago, thanks to a great friend from my Anthrozoology class at Canisius College, I was able to not only go hear Dr. Jane speak, but to spend some time with her afterwards.

I had never heard her speak before in person, and while YouTube is good, it certainly doesn't convey the energy before, during, and after hearing Jane speak in person. I too, do public speaking for my job and coach others on technique. I watched amazed how Jane spoke, without notes, without a single "um", amazing eye contact with the audience and of course, being an Irish storyteller myself, I loved her stories.

After she was done and the crowd began to leave, I walked up to the stage area and was greeted by a few members of her team and, along with a few others, was brought back to a small room nearby.

We waited anxiously and I really had no idea what to expect. She and I had emailed briefly about One Health a few months before, but I was sure she wouldn't remember.

She came into the room alone, holding a few bags, not tired like I would be after speaking for 90 minutes, but full of energy and so happy to see everyone. A few people started to just spontaneously cry. Dr. Jane appeared to look right at me, but I quickly realized she was using her x-ray vision to spot the young girl behind me who was vibrating with excitement. Jane went right to her, and it was amazing to watch. After spending some good time with a few others, she came to me.

I had gotten hold of an original National Geographic with her on the cover and had it tucked under my arm. I told her what a joy it was to meet her and reminded her that we had emailed. She asked for my name and said, "I remember".

I told her that as a veterinarian and new anthrozoologist, I so wanted to bring her messaging to the veterinary practitioner world, and to someday have even just a small amount of the impact that she has had. She took me by the hand, looked me in the eye and said, "I know you will".

It hit me hard, and I think my brain went blank. She then said, "Would you like me to sign that?", pointing to the Nat Geo I had tucked away. I fumbled out a "yes"!

It didn't hit me fully until I was on my way back to the hotel. I was so appreciative and honored. Since then, I have maintained a strong passion to do what Jane said she knew I could do and will continue to both professionally and personally. I was sent on a mission!

Thank you, Dr. Jane for what you do and what you do... does.

PATRICK FLYNN – Veterinarian

Nine Decades of Advocacy, Inspiration, and Kindness

— Patti Ragan —

While there are so many reasons to celebrate Dr. Jane Goodall's 90th birthday, it is most fitting to commend and applaud her work of advocacy to help animals live better lives, and as a result, help people live better lives. Jane said, "Only if we understand, will we care. Only if we care, will we help. *Only if we help, shall all be saved.*" Her unfaltering energy and determination to educate and elucidate hundreds of thousands of people about chimpanzee behavior, animal welfare, and environmental protection have enlightened *world thought* so that people *do* care and begin to help.

In the book *From Elephants to Mice...* the author wrote that in 1990, Jane toured a medical research laboratory in New York where chimps were used for invasive biomedical experiments. When she saw infant chimpanzees languishing alone in small cages (separated from their mothers), she suggested to the head veterinarian that he start a type of "chimpanzee kindergarten" for the infants where they could play together to experience companionship and have puzzles and toys to keep them challenged and stimulated.

After her visit, the veterinarian did start such a program as Jane suggested... and *decades later*, those first three infants she saw in the research lab finally arrived (now in their 30s) for permanent sanctuary care at our Florida sanctuary, the Center for Great Apes. In 2023, when Jane paid us a surprise visit at the sanctuary, she had a lovely reunion with those first babies she met in the NY lab – Sabina, Josh, and Ewok.

In addition to her advocacy to help chimpanzees and other animals,

Knuckles, a chimpanzee with cerebral palsy, lived at the Center for Great Apes, founded by Patti Ragan. Jane and Knuckles met multiple times over the years.

Image Source: Jo-Anne McArthur / NEAVS / We Animals Media

Jane's compassion and care for people is honest and heartfelt. During the COVID pandemic, she was grounded in England and unable to travel for her lectures. In late 2021, she heard that our very special chimpanzee (Knuckles) who had cerebral palsy had died during a seizure. Knuckles was born at a Hollywood trainer's compound and was sent to the sanctuary at age two. He was an iconic resident at the Center for Great Apes as he overcame many physical challenges during the 20+ years he lived here. Jane had met Knuckles in 2005 on an earlier visit to our sanctuary.

When she heard the news of Knuckles' passing, she emailed me to offer a special Zoom Call meeting with all our caregivers to console them. During that hour-long video call, she told us of her great love for David Greybeard and how difficult it was for her when he died. That call with Jane Goodall was both inspirational and comforting to all our staff, and I will always be grateful for her loving compassion and remarkable kindness.

The Center for Great Apes, now in its 30th year, has rescued and given lifetime care to more than 85 great apes (chimpanzees and orangutans). This sanctuary exists with the inspiration of Jane's work, wisdom, and compassion.

PATTI RAGAN – Founder Center for Great Apes

centerforgreatapes.org

The Impossible Ubiquity of a Heroic Difference-Maker

— Peter Biro —

Turning 90, what a blessing, and a what a nice opportunity to thank you for who you are !

To know Jane is to live in a state of constant wonder, amazement, exaltation, urgency and hopefulness. She possesses so many extraordinary qualities. These include her compassion for all creatures great and small and for the earth itself, her infallible moral sense, her exquisitely principled pragmatism, her unquenchable curiosity and thirst for knowledge and understanding about the natural world, her basic decency and, given that she is one of the great rock stars of our time, her genuine modesty.

But perhaps Jane's most astonishing quality, which remains true even as she approaches her 90th birthday, is her omnipresence, by which I do not refer to her media ubiquity, but to her literal, in-person, presence seemingly everywhere almost all of the time! This can only be attributable to her resilience, stamina and determination to use every available moment, every opportunity to bend every ear, enlighten every mind, and touch every heart. That is why she travels almost three hundred days out of every three hundred and sixty-five and why she is present by way of video conference, telephone and other electronic media the rest of the time.

Even in those precious few spaces which are ostensibly consecrated for rest, contemplation and self-care and recharging, Jane is unable – or unwilling – to pause rather than to continue her relentless campaign of public interest advocacy in the cause and service of the chimpanzees, the forests, the earth, and all of humanity. She is possessed of – indeed, defined by – what Hannah Arendt called "amor mundi" – love of the world. That turns out to be a rather demanding, full-time commitment!

Recently, when she paid a visit to my cottage for the express purpose of taking refuge in the forest on the lake and finding respite from the furious pace of her permanent world tour, she could not resist the urge to instead write letters to supporters, conduct video conferences and record video messages to various stakeholders, donors and policy-makers. She even recorded one especially moving video message for a total stranger who, Jane had come to learn, was living out the final days of her life. Jane concluded that message with the following words: "I look forward to my own similar journey and I can't wait to meet you on the other side". I am told by the husband, now widower, of that woman – who died one week later by MAID (medical assistance in dying) – that Jane's message to his wife made all the difference in the world.

That is what Jane does. She makes all the difference in the world.

PETER BIRO – Founder of Section 1, a Senior Fellow of Massey College, and Chair Emeritus and Past Chair of the Jane Goodall Institute, Global

Jane's Greatest Discovery

— Peter Singer —

If you search the internet for "What did Jane Goodall discover?" you are likely to be taken to a page on the website of the National Geographic Society where you will read: "Jane Goodall was the first person to observe chimpanzees creating and using tools – a trait that, at that time, was thought to be distinctly human." But for me, the most important thing that Jane did was to narrow the gulf that philosophers, theologians and others have dug between us and animals.

For centuries, especially in Western thought, we have categorized all non-humans, whether chimpanzees or snails, as "animals" and thought of ourselves as something quite distinct from animals. Some thinkers told us that we are a special divine creation, made in the image of God, and with an immortal soul. Descartes went so far as to say that all animals are automata, not even conscious.

Jane broke scientific conventions by giving names to the chimpanzees she was observing, and showing us that they are involved in complex social relationships and have individual personalities. In reading Jane's work, we recognize that chimpanzees, like us, express positive feelings with a hug or a pat, and when angry, can be aggressive and violent.

It is of course true that Jane also discovered that chimpanzees shape twigs to use as tools in catching termites, but the importance of this discovery lies in what it tells us about the ability of nonhuman animals to plan ahead. To select a straight twig and strip off its leaves does not count as making a tool unless you are planning to use it as one.

The statement you find on the internet gets it wrong. To claim that Jane was "the first person to observe chimpanzees creating and using tools" is to deny the chimpanzee personhood that Jane has so painstakingly observed and persuasively described. The first person to observe chimpanzees creating and using tools was another chimpanzee. Jane was merely the first human to observe it.

By narrowing the gulf that we had conjured up between humans and animals, Jane has provided a strong scientific basis for those who advocate for respecting the interests of nonhuman animals. She has, of course, used that basis superbly herself, but it is even more important that she has changed forever the way we think about ourselves and other animals

PETER SINGER – Professor of bioethics at Princeton University, is the author of Animal Liberation, first published in 1975, which led Jane Goodall to go vegetarian. In 2023, he published a fully updated version, Animal Liberation Now

petersinger.info

Jane Goodall's Secret Weapon

— Phee Boon Kang —

Turning 90, what a blessing, and a what a nice opportunity to thank you for who you are !

One could attribute the incredible tapestry of Jane Goodall's fabulous life to the multitude of innate gifts she possesses beyond charisma, in particular, her boundless curiosity, deep compassion and empathy for both human and animal emotions, extraordinary willpower and her iron constitution. She has fearlessly and tirelessly embraced the challenges of constant global jet lag to mobilize millions around the causes she cherishes. However, her most pivotal and remarkable gift might be her incredible power of observation.

Jane has deftly honed and utilized this gift to make groundbreaking scientific discoveries and establish her unique global educational youth movement. Her career was initially jumpstarted by her seminal discovery of wild chimpanzee behaviors observed at close range in Gombe Stream National Park, Tanzania, where she patiently cultivated the trust of the wise old chimpanzee "David Greybeard." Recognizing that her impact would be limited if she remained confined to the forest, she resolved to extend her efforts beyond and effect change within human society itself.

Since then, Jane has advocated for important emerging world issues with urgency and conviction. Beyond animal welfare, chimpanzee protection, and community-led conservation, she has also assumed a leading role in countering emergent threats to our planet, such as environmental conservation, sustainability and climate change.

Her extensive travels spanning decades have not only allowed her to inspire countless audiences but have also facilitated the recruitment of numerous "Friends of Jane" from all walks of life. With astute discernment and instinct honed through decades of observing both chimpanzees and humans, including interactions with world leaders, she has brought aboard influential allies. Her acute observation skills and attention to detail come into play when she unfailingly composes warm thank-you notes to express her genuine gratitude.

Jane attributes her own success to the countless friends she has garnered. She often likens them to individual feathers comprising the powerful wings of an eagle, symbolizing their collective impact on her cause. Furthermore, she affectionately refers to her close male friends from different corners of the world as supportive "brothers," to complement her actual younger sister, Judy.

Jane has illuminated the essential need for the human race to find a way to sustainably and ethically co-exist with our animal neighbors. Her unwavering optimism and trademark compassion have inspired many to contribute to making the world a better place through their own individual, daily actions. Given the existential urgency for our species to course-correct itself, Jane, a vital nonagenarian, continues to steadfastly and incessantly dedicate herself to her life's work, unquestionably placing her among the greatest individuals in history.

PHEE BOON KANG – Chair Emeritus and Founding Member, JGI Global. Honorary Chair and Founding Member, JGI Taiwan

linkedin.com/in/boonkang/

Tribute to Jane Goodall

– Philip Lymbery –

S ome of my most treasured moments have been when talking with Jane. Her insightful, plain talking, delivered without any ums or errs, and always with gentleness and humility, has always seemed the embodiment of strength and compassion.

Jane has always been such an unwavering voice for a better future for all life on this lonely planet. Seeing animals for what they truly are – individuals, with their own personalities, their own wants, needs, likes and dislikes.

Her empathy for animals has always been crystal clear. Hats off to her during early days as a researcher, having the courage to go against the grain of scientific convention at the time, which was very much to deny the emotions of animals. Ignoring those at Cambridge who specifically told her not to talk about chimpanzees having personalities, minds or emotions.

That to me is an embodiment of compassion in action. A stoical commitment to speaking truth to power through deeds fueled by empathy and intuition.

I had the enormous privilege of working with Jane not so long ago at the Museum of Natural Sciences in Brussels. It was a symposium on the future of food.

Jane gave the opening remarks, I did the closing ones.

Inspired by her words, I drew on Jane's take that human intellect has enabled a rather weak and unexceptional species of prehistoric ape to evolve into what *Homo sapiens* have become now: "self-appointed masters of the world".

I remembered how she draws a distinction between intellect and intelligence; after all, "an intelligent animal would not destroy its only home".

Standing amongst the dinosaur exhibits of the museum, I was struck by how fragile life on Earth really is and how human intellect could well land us with the same fate.

How the war on nature embodied by factory farming, with its wild-life-killing pesticides, cages and deforestation, is threatening the future for us all.

With time running out to save our only home, I take great comfort from Jane's thoughtful analysis that perhaps it's time for our species to evolve once more: to turn intellect into intelligence.

As Jane has described it, something that will require our head and heart to work together: combining intellect with compassion.

Thank you, dear Jane, for your profound and inspirational vision of a life-saving evolution. For showing how compassion is an essential part of a new intelligence. A new era. A new hope.

PHILIP LYMBERY – Compassion in World Farming.

ciwf.org.uk

Jane: ... Living in the Question

— Reed Oppenheimer —

Turning 90, what a blessing, and a what a nice opportunity to thank you for who you are !

To know Jane the way that I, and many of you do, is to live in a state of constant enquiry; "What more can I do?" Whether adding sugar to your coffee cup before adding the coffee so that you do not use an environmentally harmful swizzle stick, starting a regenerative farming educational center

Jane and Reed making a 'We Want You' sign.

Image Source: Reed Oppenheimer

in Uganda to demonstrate how humans and threatened species can live in harmony, flourishing together on this precious thing we call earth, or simply stopping along a busy highway to help a turtle cross the road, it is always that question that matters.

Years ago, Jane and I spent the better part of an evening, and a proportionate amount of a bottle of scotch whiskey, proposing the question, "Is it more valuable to save the world or to make a dog wag his tail?" We both concluded that they were equal. It was living in the question, "What more can I do?" that was so important.

The Universe constantly reminds me of the Power of Jane. Last month in Tanzania I met Mulubi Mulanda, a Ugandan who had come to be certi-

Planting a tree

Image Source: Reed Oppenheimer

fied in Restorative Agriculture. He had met Jane 30 years ago in a refugee camp and had been inspired to dedicate the rest of his life to teaching others how to heal the planet and themselves, reaching thousands of families living with little hope. The Power of Jane was forever etched in his soul.

I regularly meet middle aged adults demonstrating marvellous commitment with their lives only to find that they were inspired three decades ago by the powerful messages of personal responsibility enshrined in the Roots & Shoots program. Many had listened to some of the songs and messages that Jane and I had recorded when they were children to inspire good stewardship of the planet.

To know Jane, is to live in the Question. Just a few months ago I was inspired to plant a native and threatened Koa tree in a sacred forest on the Island of Hawaii in honor of Jane. For decades to come, indigenous people and others will come, touch the tree, feel its blessing, and live in The Question, "What more can I do?"

Jane

Love is truly countless as the sands upon the shore.

Commitment has the power of an Ocean when it roars.

Both ever changing, neither in control,

They create an endless shoreline and define each other's role.

May your purpose never waver to let creation be.

Man and beast in harmony; Young roots growing to be free.

And may I ever walk beside you where your vision's dear to me.

May we always find each other where the shoreline meets the sea.

Love,
Reed

REED OPPENHEIMER – Chairman/CEO Reed Jules Oppenheimer Foundation

youtube.com/watch?v=QR_YBnYfjP4

(Reed planting a Koa tree for Jane in a sacred forest on Hawaii, where his foundation tries to re-establish indigenous forests on which more than a thousand endangered species depend for survival.)

How You Inspire Action

— Rhett Butler —

One of the earliest nature books I read on my own was *My Life with the Chimpanzees*. This book not only amplified my interest in animals and tropical forests but also instilled in me a dream – a dream to one day be a part of that wondrous world. Driven by this passion, I immersed myself in every natural history book I could find and constantly urged my parents to take me to visit the rainforest. It wasn't long before I seized that opportunity, and from those initial experiences, the foundation for Mongabay took shape.

Long before I started Mongabay, however, it was your work, Jane, that became my beacon. Your unwavering commitment to advocating for animals, paired with your relentless drive to challenge deep-seated assumptions about other species, deeply resonated with me.

I never imagined that I would someday meet you, let alone become friends. In many ways, I've always felt that we are kindred spirits. I quit my job to pursue Mongabay full-time at the same age that you set off for Gombe, also without formal training, but fuelled by an insatiable curiosity, a profound compassion for other species, and an understanding of the significance of fostering a bond between humanity and nature. Our journeys, though distinct, have serendipity as a common thread, which, in time, united our paths.

Your monumental contributions to conservation and your unyielding compassion for all beings extend far beyond merely inspiring me to establish Mongabay. You've imparted a profound lesson – that every entity, be it human, plant, or animal, holds intrinsic value and plays a pivotal role in the grand tapestry of life. Your life's work serves as a reminder of the indispensability of hope. For without it, what do we truly have? You've exemplified the transformative power of advocacy, emphasizing that we must vociferously champion the causes close to our hearts.

Thank you, Jane, for the boundless inspiration and for your immeasurable impact on our world. Wishing you a joyous 90th birthday!

RHETT BUTLER – Founder of Mongabay, a conservation media service that delivers news and inspiration from Nature's frontline

mongabay.com

Jane Goodall as Scientist

– Richard Wrangham –

Dear Jane,

As you will doubtless remember, in *Brazzaville Beach*, published in 1990, the novelist William Boyd imagined the recollections of Dr. Hope Clearwater, who as a young ethologist had studied chimpanzees in a fictional National Park, Grosso Arvore. The chimpanzees were observed partly with the help of an Artificial Feeding Area, and at first they had appeared to be entirely peaceful. But then "[f]or some unknown reason, a small group of chimpanzees had broken away from the main unit" and had gone south. Signs of violence followed. First there was infanticide and cannibalism, then boundary patrols, then deaths of adults from violent coalitionary attacks. Hope Clearwater reported all this at the time that it happened, as you did. Later in life she lives on a tropical beach next to the ocean, as you do.

All this was an accurate account of true events in Gombe. It was your life. Yet astonishingly, Boyd's acknowledgment merely included you as one of a dozen people who had been "helpful and instructive." "helpful and instructive"? How about "the vital inspiration for an almost plagiaristic account of an extraordinary scientific story"?

But if Boyd's novel was less fictional than readers might have realized, at least it gave a perfect portrait of your relationship to science. When Clearwater prepared to publish she was confronted by a senior scientist, Eugene Mallabar. "You are jumping to conclusions," said Mallabar, who had committed himself to the idea of chimpanzees as inherently peaceful. "Bad science, Hope. Whatever you may think is happening is wrong." He told her to stop what she was doing. "Leave the interpretation to me." Clearwater was unmoved. She published her findings at once.

Boyd got it right. As a scientist you have always been resolute. You are honest about what you have documented, and you know what's important to document. In Gombe you advised us students always to keep theory in its place. Of course ethological descriptions can never be free of theory, because we need theory to tell us where to pay attention. But what I learned from you was how much one gains from being consciously open-minded. One might see something that doesn't fit into established theory, or that doesn't conform to the boxes on one's check sheet. Make written notes! Think about what it means! And all of a sudden you'll be a pioneer, as you have endlessly been and as Boyd fully recognized you are, because in your hands science has always been as demandingly honest and revelatory as it should be.

RICHARD WRANGHAM – Graduate student at Gombe, 1970-1973. Research Professor in Biological Anthropology, Department of Human Evolutionary Biology, Harvard University

kibalechimpanzees.wordpress.com

Jane's Tunnel Vision

– Rick Quinn –

The physical and mental stimulation was about to extend beyond the intended canine participants. With one hand covering the microphone and a mischievous look in her eyes, Jane asked me if she might do something incredibly silly.

Jane was to be awarded an honorary Doctor of Laws at Western University in London, Ontario, Canada. Jane, Susana, and the JGI Canada team would be guests at our rural home. The advice on how to ensure Jane herself enjoyed a fundraiser contemplated for the evening before was simple: include dogs. Six dogs and their handlers would compete in a canine agility event – a sport where you guide your dog through a pre-set obstacle course in a race for both time and accuracy. A white peaked wedding tent was erected outdoors to protect the guests – six dogs and one hundred people – from the chilly springtime elements. One end of the tent housed the agility course, complete with weave poles, a seesaw, a long tunnel, a tire jump, platforms, and jump bars.

Bartenders mixed their signature drinks: the 'David Greybeard' and 'Gombe Stream' cocktails. Three local chefs tempted guests with delicious appetizers, created in front of them. All present were impressed with the six professional handler-dog teams as they effortlessly and flawlessly worked together to complete the agility course. Six guests, randomly selected and assigned a dog, subsequently completed the same course amidst a chorus of well-intentioned laughter, receiving ribbons for their new found dog-handling skill.

That's when it happened. Unscripted, Jane calmly asked for a group of volunteers. On hands and knees, Jane proceeded to crawl through the tunnel, followed closely by her band of volunteers, emerged from the other end and headed for the tire jump. Most observers held their breath. Somewhere between Jane's squeezing through the tire jump and walking along the seesaw, I overheard the visibly blanched University Provost's hope for Jane to emerge intact and available for the next day's convocation. She related that it was the first time in the university's 140-year history that convocation was to be held for a single degree recipient – and Jane really needed to be there.

Less than twelve hours later, a reserved Jane Goodall would deliver a flawless convocation address, to an adoring capacity audience. I sat in awe, several feet away on the stage, with her notes for the address on my lap. I would soon "hood" one of the most powerful voices on the planet.

RICK QUINN – DVM, DVSc, Diplomate ACVO. Founding Director, Docs4GreatApes. Director, the Jane Goodall Institute (Canada). Director, the Jane Goodall Institute (Global)

A Personal Tribute to Jane Goodall on the Occasion of her 90th Birthday

– Rob Muller –

I first met Jane in 2007, as part of a JGI-US strategic planning project where I was to work with board, staff, stakeholders, and of course Jane to craft a plan that bridged chimp and primate research and protection, community centered development, and youth environmental action. We met in London; it was spring. We went for walks and talked about Jane's vision, how to build a global movement to sustain her legacy, and also world affairs and aspirations. It was the start of a long link to Jane and the work of JGI sparked in childhood.

Public figures come and go. Fads fade fast; our collective attention span gets shorter and shorter. But Jane stays on. Evidence a recent speaking engagement at the 3,600-seat landmark Chicago theatre. Sold out, standing room only, crowd gathered at the stage door waiting to catch a glimpse of the peripatetic Jane Goodall.

Why has Jane and her work captured the imaginations of people like me and spanned generations and decades? It is in large part due to her message of hope in response to today's general malaise and climate insecurity. It's also in the values that underlie that message. It is her conviction that individual people matter – no matter who or where they are. Kindness, generosity and curiosity. That a meaningful life is one spent in service to others. Certainly large-scale societal actions are essential to saving the planet. But that seismic shift in policy and behavior rests in each individual. Jane's tactics – of inclusion, listening and reaching out instead of lashing out – has the power to bring people together.

A meeting of representatives of JGI Chapters around the world in Vienna, 2022. This was the first global meeting in many years due to COVID interruption.

Image Source: Marko Zlousic

Jane's messages resonate across generations – from my parents who came of age in the 1940s, to me, to my young adult children, and (we hope) to our eventual grandchildren. Recently, I've joined the Jane Goodall Institute Global Board of Directors and currently serve as co-chair. I've had the privilege of observing Jane's impact and that of the institutions that carry her name. We need more Jane Goodalls. We need to make sure that the global movement of Jane Goodall Institutes grows and carries forward for decades to come.

ROB MULLER – Co-chair of the Jane Goodall Institute Global Board of Directors

The Symphony of Hope

— Sir Robert (Bertie) Eden —

I have been fortunate to have shared with Jane several events on several different continents. Some have made me cry laughing with joy some have made me weep with sadness, and others have driven me to despair where there has also been hope.

For one event we were in Vienna staying at the British Embassy, as Jane is a DBE she has this privilege. It was very pleasant, sort of like being at a rich Aunt's place that you see one night for a formal dinner during your stay.

It was a small event at the University of Music & Performing Arts and potential donors, scientists, politicians and friends had been invited by the JGI Austria team.

It was very distinguished and quiet but the arrival of Jane put everyone at ease and smiles broke out and a certain gaiety was in the air. (The Jane Energy).

We all took our seats opposite the orchestra. Gudrun, the then Executive Director of JGI Austria, stood in front of the orchestra to introduce the event. I was sitting on one side of Jane and Walter Inmann on the other. Gudrun is one of those people whose beautiful soul shines clearly from herself. I still to this day don't know how she did it but she delivered a speech with floods of tears rolling down her cheeks, intermittently giggling and smiling, as she described her work for JGI and her love of Jane.

That would have been enough emotion for me for one evening and then the orchestra started playing.

The music was blissful and breathtaking. It was symphony No45 by Haydn. The 2nd movement is dreamy and brings you to a deep meditative state. The final movement snaps you out of it into a fast tempo and then suddenly breaks off, at this point each instrument group of musicians has a small solo at the end of which they rise and leave whilst the others still play. I was not prepared for this as I didn't know the piece. I looked at Jane inquisitively and through her concentration she nodded and smiled. The end is soft with a single violinist on stage finishing the piece.

With the applause the orchestra returned to their seats and Jane got up to speak. She delivered a talk on the destruction of habitats by humans that have devastated species around the globe.

Her description of many of these destructive events was poetically terrible and many tears flowed. Jane related species extinction to the orchestra disappearing but reminded us that there is hope as the orchestra came back. I believe all of us left that evening with a beautiful hope in our hearts and minds.

SIR ROBERT (BERTIE) EDEN – Wine producer and Chair of Governance Committee JGIG Global Board

A Duet of Chimps and Wolves

– Ron Kagan –

I first saw Jane's advocacy efforts 35 years ago. Two chimpanzee drowning incidents at the Detroit Zoo, one involving a dramatic rescue by a visitor, led Jane to publicly challenge that zoo's seeming absence of compassion, its policies and its dangerous facilities.

As a member of the American Zoo association's chimpanzee management group, I had been working for some time to get U.S. zoos to stop building water moats as containment for chimpanzees. There had been occasional reports of drownings for years. After extensive research I learned that almost half of all the world's zoos that used water moats, had one or more fatal (or near fatal) drowning. An astonishing statistic, except for the fact that chimps can't swim. Containment shouldn't kill.

Ironically, a few years later I would go on to head that zoo, spending millions to remove the chimpanzee exhibit water moats. Ultimately new zoo association guidelines discouraged using water moats.

A decade later (2004) our zoo made the decision to end the practice of keeping elephants. After years of major changes to care practices and facilities, we concluded that it could likely never be "enough" for them to be able to thrive. Our realization was slow, perhaps by decades. Sometimes experts in scientific fields have difficulty overcoming hubris.

It was not a popular, or acceptable decision for many in the zoo world. So, I asked Jane and others like Cynthia Moss to help with those who opposed, and for a time blocked, moving the animals to a sanctuary. Jane's words mattered. Hardly the first time Jane advocated for zoos to be champions not just of conservation and education... but of animal welfare. Her impact on the zoo world over the decades on many issues has been enormous.

A few years later Jane was recognized by a Canadian Animal Welfare organization. Since I had been the previous year's recipient it was customary (and an honor) to be the one introducing her at the University. The night before, due to a television broadcast request, I was asked to give "after thoughts", instead of an introduction. Really, follow Jane Goodall?

Jane began as always with chimp calls, stories of her U.N. peace work, stories about individual chimpanzees, and her life at Gombe. For an hour everyone was mesmerized. Naturally.

Then came I. I apologized for being in the way of those who wanted Jane to sign their books, or to meet their children... or those who might need the restrooms. I apologized further that while I too have been talking with animals for decades (notably with wolves), I don't make animal calls to, or in front of, people. After a few brief remarks encouraging more public dialogue about zoo animal welfare and ethics, I turned to sit down. Jane jumped back to the dais, literally grabbed me, and pronounced that we would not let Ron go without his Wolf howling. I froze. I protested. But, Jane held me tightly and started "pant-hooting" ... I sighed. Dropping my head back I joined her by howling. Soon, several in the huge audience howled back. More than a duet!

RON KAGAN – Animal advocate, Compassionate Conservationist, Zoologist

A Heartfelt Tribute to Jane Goodall

— Rooney Mara and Joaquin Phoenix —

What makes Dr. Goodall truly special to us is her ability to bridge the gap between science and empathy. Her profound connection with animals and her tireless efforts to create a harmonious coexistence between humans and the natural world speak to a deeper understanding of our shared planet. Her work serves as a poignant reminder that empathy and compassion are essential tools in our quest for a sustainable and harmonious future.

Jane, thank you for the profound inspiration you have bestowed upon our families. Thank you for your intrepid bravery, unwavering optimism, tireless dedication; and for your earnest, motivating credence that our species is capable of better. The world is indebted to you.

ROONEY MARA and JOAQUIN PHOENIX – actors and activists

A Tribute to Jane Goodall

— Satish Kumar —

J ane Goodall is my friend, a mentor and an eternal inspiration. She is a gracious gift from the universe to the world. She is one woman movement in the service of our precious planet Earth and its people.

My close encounter with Jane took place in unusual circumstances. We were participating in the launch of the Arc of Hope and a celebration of Earth Charter at Shelburne Farm in the state of Vermont on the 9th of September 2001. We were joined by great musician Paul Winter and environmentalist Steven Rockefeller. It was a joyful occasion. On the 10th of September Jane and I travelled together to New York. We were to fly home on the evening of the 11th of September. But on that fateful morning of 9/11, the terrible terrorist attack took place on the Twin Towers of the World Trade Centre and on the Pentagon in Washington DC. All together 3000 people were killed and of course not only our flight to London but all the flights departing from America were cancelled.

Thus Jane and I remained in New York for about a week, meeting the families of the victims and terrorised citizens of New York. We were two messengers of peace, having spent some happy time together in Vermont, now in contrast we had the sad task of consoling people in New York and participating in the search for peace and justice while the US government was preparing a long war against terrorism.

Friendship made in difficult times is strong and long lasting. Ever since Jane and I have worked together to bring hope to humanity. She has been a great inspiration to me in establishing Schumacher College in Devon, a college of hope. Jane brings the spirit of no ordinary hope but a profound sense of "Active Hope!" In order to bring hope to the world Jane has devoted herself 100% to the service of human world as well as to the more than human world. She is a tireless traveler, a superb speaker and a brilliant writer. She has become a living legend. Although we are celebrating her 90th birthday, to our amazement she is eternally youthful. She is intuitive, imaginative and lucid in her ability to communicate and transform people who come in contact with her.

I congratulate Jane for her courage and dedication, her love and commitment and her determination and willingness to help humanity to make peace with the Earth and with each other.

SATISH KUMAR – Founder, Schumacher College and Editor Emeritus, Resurgence & Ecologist

What a Privilege

– Shadrack Mkolle Kamenya –

I feel very fortunate to have known this lady named Jane Goodall, am also very glad that she chose to come to Tanzania and specifically coming to Kigoma Region which was a golden chance for us, I do not want to be selfish, but I thank God for that very choice. The nation and people have witnessed Jane changing from being Jane Goodall known to few people to Jane Goodall of millions of people, nationalistic figure well known all over the world, and wherever people are talking about her, her work, her projects scattered in different places in the world, then Tanzania, and Gombe in Kigoma is also mentioned therein. The topic she first came to research on could have been studied in other sites where the chimpanzees are found, but she chose coming to my homeland, what a privilege for us. Doctor Jane Goodall has helped increasing visibility of Kigoma and Gombe in global map and has made them popular to many people around the world. There are many visitors coming to Kigoma and Gombe in particular because of Jane Goodall's lectures, books, articles that have promoted the famous Gombe chimpanzees for the past 60+ years. Jane Goodall was nominated as the UN Messenger of Peace for us she is more than that, she has been the Messenger for chimpanzees of Gombe, Kigoma and Tanzania environment around the world for many years. Many if given opportunity would witness similar issues about their places for various reasons.

What a privilege to be around somebody who makes use of time she gets on this planet to do the best she can for environment (other people, planet, and biodiversity) and very little for herself. Many do lots for themselves and very little for others, Jane Goodall not that kind of a person. I just

Some Gombe researchers during a meeting of the International Primatological Society in Chicago, 2016. Seated at the left of Jane are Lilian Pintea and Shadrack. Just behind them is Ian Gilby. Anne Pusey is two seats at the right from Jane, on the second row. Standing on the last row, third from the left is Deus Mjungu.

Image Source: The Jane Goodall Institute

wonder how she learned to be content and a very giving person. Perhaps by being very close to her given the time I see her every year I have learned a glimpse of who she is, it is hard to hang around her and resist a change. She has influenced many people around the world, and she does not know how many people she has but I think unaccountable number of people could line behind her in their thinking and practices, and that gives me hope for this planet earth becoming a better place even for the next generations. I can imagine the mushrooming Roots & Shoots members around the world who some are willing and walking in Jane's footsteps.

I have seen people in my homeland when they turn age 60 plus years old, they feel too old to work and they declare being old and begin shunning away from work, or when reach retirement ages, few reach age 70 years working but rarely you find people working at Jane Goodall's age and keeping up the pace. Many wonder where she gets her energy and strength in doing what she does? Jane Goodall has her answers to this question, and some of us can make our guesses, perhaps Jane's enthusiasm, love of what she does and continued effort, others like me may conclude that Jane is the most blessed by God the Creator of all beings with energy, time, wisdom because she is God's great steward who needs to go around the world proclaiming the better care, love of the created beings and our environment for better tomorrow. Wherever that comes from, I wish and pray that the work continues, more people reached and changed for the better tomorrow. Thank you so much Doctor Jane Goodall for all that you have done and continue doing and truly for being a blessing for many people on this planet earth.

SHADRACK MKOLLE KAMENYA – Director of Conservation Sciences, JGI Tanzania

Jane Goodall: The Lighthouse of Love

— Shweta Naik —

The intricate tapestry of human history has its share of luminous figures, those who are able to carefully blend compassion with a sprinkle of humor and in Jane's case, perhaps a dash of some good vintage whiskey. She's a beacon whose presence radiates the warmth of love, the steadfastness of guidance, and the profound interconnectedness of all living beings; she is truly the guiding light of a lighthouse, neatly wrapped up in a bit of whimsy!

My personal journey with Jane unveiled to me, the depth of her impact, a transformative experience that truly mirrors the essence of a lighthouse; standing steady on a bedrock of hope, doing the job of guiding weary hearts and minds through life's unpredictable waters.

Jane's presence came as a lifeline of hope, during a time when I was wearily waddling through life, heavy under the weight of the trauma that accompanied my daughter Anaiah's open-heart surgery. Fate had somehow intervened and put me right in front of Jane. Enveloped in her love on the veranda of her home in Dar es Salaam, strength found a way into my weary heart. This connection, paralleling Anaiah's healing journey, sparked my transformation, much like a lighthouse guiding a ship to safety.

As I sit down to write this, I am reminded of several anecdotes peppered across my adventurous journey with Jane, however I would like to mention one in particular. This one truly captures the essence of "Being Jane". One nondescript morning, we happened to venture into a shop in Dar es Salaam, Tanzania, with Jane. The incumbent shopkeeper's anger and irritation at an employee hung heavy in the air. Then, Jane walked in. The shopkeeper's demeanour transformed drastically. Her greeting changed the entire atmosphere, as if a serene breeze had swept through the room. The shopkeeper, once agitated, suddenly became the embodiment of calmness and gentleness. He greeted Jane with deference and humility, captivated by her presence. A brief exchange sufficed to dissolve the tension that had pervaded the environment. Janes' presence alone managed what seemed impossible—to infuse tranquillity in the room.

Another vivid memory: a call from Dr. Anthony Collins, sharing news of Jane's arm injury in Gombe. Amidst Roots & Shoots events and documentary shoots, I witnessed her resilience, even as she grappled with combing her hair. Her unwavering dedication, humility, and resilience etched an indelible mark on my heart.

Jane's journey weaves a legacy of profound love for the natural world – a legacy that extends beyond scientific discovery. Jane is not confined to the academic tower; she embodies a friend, a mentor, and a beacon of approachability. Her interactions empower individuals, instilling them with the belief that they too can champion Earth's cause. It's in her engagement that her legacy finds strength—an inspiration reminding us that profound impact blossoms from the simplest exchanges. Jane's journey also illuminates oneness—an affirmation that we're not isolated entities, but individual threads woven intricately into life's fabric. Her advocacy for environmental causes emanates from her belief in the inherent bond between all living things. Like a lighthouse's beam, her message transcends borders, fostering unity across cultures and divisions.

Much like a lighthouse—a sentinel of safety against tumultuous tides, guiding ships through perilous waters - Jane's influence navigates hearts and minds through the uncharted currents of conservation and personal transformation. Her teachings, dedication, and warmth break through the fog of ignorance, illuminating the path ahead.

Shweta Naik – Executive Director, Jane Goodall Institute India. Shweta stumbled into Jane during the darkest night of her life, and has since found her shore, where she now aspires to set up her own lighthouse; One that will guide life's lost sailors by introducing them to the work and world of Dr. Jane Goodall

Unlikely Beginnings

— Stephan Margolis —

The 1990s was a crazy decade. In April 1992, I found myself, along with thousands of other Los Angeles Police Officers, deployed to South Central Los Angeles to manage an explosive riot. In less than a year, I was re-assigned permanently to South Central to help rebuild police – community relations amidst a staggering rise in murders and gang violence. Against this landscape, it is hard to imagine...

It was late evening at the gym, where I ran into the owner, Blinky Rodriguez, a world kickboxing champion. Blinky had an evangelist commitment to negotiating a "gang truce" in Los Angeles. He more than occasionally stepped close to the legal line in his pursuit of this vision. We had many talks, and tonight was no different – except his words were immobilizing. "Tomorrow, I am having breakfast with Jane Goodall to discuss our program." I managed only one sentence – "Get me an invite, Blinkey, and all the favors I have done for you will be wiped clean."

The following day, I was at a breakfast table filled with school administrators, kickboxers, and former gang members. To my left was the eternal beauty and renowned primatologist Jane Goodall. Her ever-present quietude filled the room. Her soft, accented voice held court. My vocal cords failed me as I absorbed every word. I knew I had to speak, or this moment would be lost forever. My imagination rushed through stacks of National Geographic magazines, videos, and books, angling for something to say. Nothing in my years of policing had prepared me for this. I was terrified.

I finally introduced myself and repeated the same spiritless line I would later hear many times over decades of presentations. "Dr. Goodall, what an honor it is to meet you. I so admire your work." She was gracious, if not wholly indifferent. The event ended, and I had achieved NOTHING.

Finally, I abandoned my trepidation and announced, "Dr. Goodall, I work in a part of the city held hostage by the rise of violence. You are an ethologist. We could use your help to better understand what to do." Her face brightened, and her eyes focused on what I was saying. She asked what I proposed. I told her we would take a police car during the high time for murders and visit fresh crime scenes. As soon as the words left my lips, I realized my proposition was a bit of a macabre script.

EXT. POLICE STATION PARKING LOT – NIGHT

A uniformed SERGEANT assists JANE in positioning her POLICE BULLETPROOF VEST before getting into the Black and White patrol car. The sergeant recites several safety issues.
SERGEANT
... if the vehicle is shot at, lie down, and Detective Margolis and I will handle it.
The night erupts into flashing lights and sirens as Jane and Mr. H study man's unshackled acts against one another.

This unlikely but true event and those that have followed brought a lifetime of love, the most profound friendship, countless contributions, and, of course, adventures.

Jane, we honor you for changing our world, hearts, and future.

STEPHAN MARGOLIS – Lieutenant, Los Angeles Police Department (Retired)

"Aloha Jane and Jau'oli La Hanua from Maui"

— Steve Woodruff —

Turning 90, what a blessing, and a what a nice opportunity to thank you for who you are !

My Dear Jane

Greetings from Haiku, Maui and a very "Hau'oli La Hanau" to you on this milestone birthday! I know you don't like to be made a fuss over, but sometimes it is a good thing for each of us to pause, take time out and thank people who have impacted and made such a positive difference in our lives, or in your case the world.

We met back in Pointe Noire Congo in 2008, as you entered "Bongo House" with your arm in a sling from a Tanzanian fall caused by a loose rock. You immediately asked me and my visiting 3 sons, Miles, Brett and Glenn, to help you tighten the sling so that it was more comfortable for you. So there we all were, standing in the living room just minutes after your arrival, tugging and pulling on you to tighten your sling as you provided feedback to us on how we were doing. The rest of the day we visited in the backyard, and you asked my sons what each was doing with their lives and if they thought they were making a true difference in the world. As a result of that encounter with you... 2 of my sons redirected their college degrees and later came to Congo to work for JGI in the jungles of Conquatti. Later you asked me to join your USA and Global Boards which I did and have enjoyed (mostly) for the last 10 years. It is just a short story on how a brief encounter with you can impact and redirect the lives of so many of the people that you touch.

Later that week we hosted a gala reception for you at Bongo House and

Jane and Steve Woodruff hugging one another

Image Source: Steve Woodruff

invited all the important politicians, Industrial leaders, teachers and friends of JGI. Right after sunset, we were all gathered around the pool drinking our "sundowners" when suddenly the overloaded generator blew a fuse and the entire party was plunged into total darkness for about an hour. I nearly wet myself, but you being so comfortable with African ways ... didn't skip a beat and simply carried on visiting in total darkness as if nothing had happened! It was a wonderful and memorable evening for everyone indeed!

You personally have influenced me in many ways as I have watched you teach, lead and inspire people for many years. Some of your biggest teachings to me have been:

- The power of reaching and motivating others through effective storytelling.

- Your ability to engage non-believers & opponents so not to alienate them but rather keep the conversation alive.

- Acknowledging that the mission is difficult and daunting, but worth it all and to never, ever give up.

- Acknowledging your celebrity/fame status but always remembering to be your own, genuine, authentic self.

- Consistently finding time to send notes of friendship or comfort to thousands of people around the world.

- Being unflappable and never getting discouraged. Seeing each sunrise as a new opportunity to make a positive difference.

- Striving to be a continuous learner and applying new technology as an opportunity for advanced problem-solving.

- Focusing on the new generation who are not yet set in their ways, more receptive to think differently, highly motivated, and willing to change their behaviors.

Well again a very happy birthday to you! Whether you call it 9/10 of a century, 90 years, 1,080 months, 32,850 days or 788,400 hours ... that is a heck of a lot of adventures, teaching, friendships, and making a positive difference in the lives of so many people, plants and animals! We all feel blessed being able to call you our dear friend! Sending you "tons of love and Aloha" from Haiku, Maui and hoping that you have a "real hoot" on this very special day!

Aloha Nui Loa

STEVE WOODRUFF – Member of the Board of the Jane Goodall Institute Global

Celebrating Jane

— Susana B. Name —

J ane, let's celebrate you. At 90, 120, 80, it doesn't matter. It is you and all you bring to the world just by being you that we all love to celebrate. So challenging to write about one story or one memory because there are so many over the almost 2 decades I have had the honor and blessing of joining your journey.

You are always an inspiration and a role model to me, Alex, Christian and our extended family and friends. Just a few weeks ago you sent one of your out of the blue interesting emails asking, "what animal would I say you resemble". A difficult question but beautiful to think of you in that form, in the form of an animal, what is so dear to you. A gazelle, I said after bringing Alex to the challenge. The reason, the gazelle is so graceful, and beautiful, adaptable and so capable of outsmarting the predators but as Tony said, that described you but not enough. There are some more attributes that you have, and you so generously share with all of us. Your wisdom Jane, and your generosity with your time and true commitment to make life better for this planet is a constant source of inspiration to work and invest time in making things better. Ever since our first phone conversation when you changed the scope of my duties to just help you and Mary with the overall agenda around the world, little did I know what I was signing in for. Little did I know I would embark in such fun, fulfilling job/friendship if I may.

Happy 90th birthday! and a very big Thank you for being you and being such an important part of my life.

SUSANA B. NAME – Vice President Founder Relations, Partnerships & Special Projects, Office of the Founder – Global, the Jane Goodall Institute

Disruptive Jane

— Susana Pataro —

It was November of 2013, and I was beginning to go through my life as a retired diplomat after 40 years of service. My professional life has taken me to various countries, international organizations, and ministerial services: France, Italy, Greece, UNESCO, Nigeria, and West and Central Africa. I admit I had decent material to begin to string together some memories. However, a disruptive event in my life turned any project of a calm retirement upside down.

Some years before, while visiting Paris, a friend took me to listen to a personal presentation of Jane Goodall at a Theatre on the Grands Boulevards. Her life and especially her research with the Gombe chimpanzees were familiar to me since I was an adolescent and saw her on the cover of *National Geographic*.

After her presentation to a warm audience, while she dedicated *James and Other Apes* to me, written and illustrated by James Mollison, she told me that she would visit Argentina for the first time later this year as it happened.

When Jane learned of my upcoming retirement from diplomatic life, she invited me to join the Global Board of Directors that had just been created to represent the Latin American region. She thought with good judgment that my multilateral diplomatic experience, my life in Africa, and my already-known passion for chimpanzees were suitable bases to join the adventure that meant helping to build the institutional structure to perpetuate her legacy in the world.

My affinity has always been with the animal world, especially with great apes, and my familiarity with the African heritage led me directly to the dire situation of the great apes. It was the year 2001 when UN Secretary Kofi Annan launched the Great Apes Survival Partnership (GRASP) initiative. I knew that our closest living relatives inhabited the most dangerous regions of the planet, both in Africa and Asia, and had not hesitated to meet them and learn about their threats as those of the human communities surrounding them.

Jane's invitation opened such a big door for me that I had to hold on wherever I could so the hurricane winds would not carry me away.

When I began to experience on me the expectations of people towards Jane, I understood that she had given the world light and hope and the tremendous responsibility lying on the shoulders of those working close to her. It was possible to relate differently with that part of us, the non-human animal world from which we have separated ourselves in mind and heart.

Jane has been a sweet disruptive force in science, philosophy, and conservation. For the good. The same happened with my life.

I feel blessed to be amid this storm and to be able to contribute with my daily personal choices and advocacy work so that apes live better and that we respect their rights to live with dignity on a planet that belongs to them as much as to us.

SUSANA PATARO – Chair of the Global Policy & Advocacy Committee, the Jane Goodall Institute

Chicago Encounter with Jane

— Tetsuro Matsuzawa —

It was November 1986 in Chicago, when a symposium entitled *Understanding chimpanzees* was held at the Chicago Academy of Sciences. Jane's book *Chimpanzees of Gombe: Patterns of Behavior*, had been published the previous year. To celebrate its publication, chimpanzee researchers from all over the world gathered for the first time in history. I was one of the youngest invitees, an assistant professor at the Primate Research Institute of Kyoto University. For eight years, I had been working with a young chimpanzee named Ai, teaching her graphic letters and Arabic numerals. Ai could use a keyboard and make statements such as "red/pencils/6" and "cups/blue/5". There was her spontaneous word order in which the number is answered last. The article was published in *Nature* in 1985 titled "Use of Numbers by a Chimpanzee". I did my lecture talking on the Ai project following the Gardners who taught sign language to the chimpanzee named Washoe. Jane, the guest of honor, was sitting in the front row listening to me. After I finished speaking, Jane asked the first question saying "By the way, how does Ai spend her time after her studies?" The question was not about the content of the research, but about the daily life of a chimpanzee. Almost instantly, I recognized what she meant. It was because I had already observed the lives of wild chimpanzees in Africa in February. In Bossou, Guinea, West Africa, they used stone tools to crack open oil palm nuts. I will never forget the first time I saw them living in the forest. Their jet-black fur shines in the morning sun. A female with a two-and-a-half-year-old clinging to her belly crosses the canopy overhead with ease. Jane's question was answered immediately, "When she's done studying, Ai lives with her childhood friend Akira and

Tetsuro Matsuzawa and Jane, when she received
the International Cosmos Prize in Osaka, Japan, in 2017.

Image Source: Tetsuro Matsuzawa

other fellow chimpanzees. I trained her to come and sit by herself in front of the computer only when she is studying." Jane smiled at me. It was a smile that said it all. Back in Japan, I continued conducting cognitive research while simultaneously conducting fieldwork in Africa. In the following years, I renovated Ai's playground and built a 15-meter-high climbing structure. Jane called it the "Tetsuro Tower" and introduced it in her lectures. Our research is to better understand the lives and minds of chimpanzees. I hope it will lead to better conservation of their natural habitat and welfare in captivity.

TETSURO MATSUZAWA – Primatologist, Former President of the International Primatological Society

Real Jane

— Tiong Piow "TP" Lim —

D r Jane speaks of two Janes. Iconic Jane, famous scientist who dons the covers of *Nat Geo* and inspires the world. Real Jane is the Jane who is constantly trying to catch up with Iconic Jane – to be inspiring, hopeful and high-energy all the time.

At the end of each day, Real Jane, will be knackered from all the mental and physical exertions. Imagine being Iconic Jane for 90 years!

So today, I wish to acknowledge Real Jane.

Dear Real Jane,

While the world may have come to revere Iconic Jane, it's the "Real Jane" who truly embodies the essence of inspiration. Real Jane, humble and ever striving, is the driving force behind the iconic figure. Real Jane's relentless pursuit of knowledge and unwavering commitment to environmental preservation stand as a testament to your indomitable spirit. It's in your moments of vulnerability, when you openly share the challenges you faced that the real Jane becomes an even more empowering inspiration to all. Real Jane has become a beacon of authenticity, reminding us that icons are not immune to doubts and struggles.

Real Jane's journey is a story of resilience, a narrative that resonates deeply with individuals from all walks of life. It's in your ability to admit your mistakes and learn from them that you became an approachable role model, shattering the illusion of perfection that often surrounds iconic personalities. Your humility and constant self-reflection reveal a profound wisdom that surpasses the accolades and recognition.

Real Jane's impact extends to the countless lives you have touched through JGI, Roots & Shoots and all your global initiatives. Your willingness to engage with the younger generation speaks volumes about your desire to cultivate a legacy of change-makers. It's in Real Jane's interactions, heartfelt connections with individuals and communities around the world, that your true legacy is forged.

In a world often dominated by surface-level admiration, it's Real Jane who becomes the quiet but profound source of inspiration. You are not limited to the boundaries of fame; you transcend them, reaching into the hearts of those who seek authenticity and purpose. Real Jane is every bit as iconic and inspiring as her celebrated image, if not more so, as you embody the essence of what it means to be human and to effect positive change in the world.

That's why we only drink whiskey with Real Jane.

TIONG PIOW "TP" LIM – President of Roots & Shoots Malaysia

The Best Kind of Problem ... Jane at Wild

— Vance Martin —

For years, the dates of the World Wilderness Congress never matched Jane's global nomadism. We both tried. Then, suddenly, Jane was available for two days in November, 2009 when we convened WILD9 in the Yucatan. This was positive news for us as we planned during a maelstrom of a global recession, H1N1 flu virus, drug cartels at war in the streets, and more. Jane's presence was a ray of light, even more than usual!

Then, more. Shortly after Jane's news, Sylvia Earle also confirmed her participation. Patricio Robles-Gil (my co-convenor) and I proposed that we entirely devote what was to be a three-hour, unscheduled space to these two, long-time friends, quietly powerful and globally-accomplished women. They were both enthused.

Jane and Sylvia individually addressed the 1600 delegates, then had a conversation with each other in a lounge setting, on main stage: two friends, talking to each other about the world we've created, the hope we need, and the possibilities to change our world. The session, unusual in format, had a wonderful impact. The delegates joined their discussion and there was a palpable sense of a community of purpose, and of respect, love and hope.

As Jane neared the end of her address, I slipped out of the main hall and strode quickly down the deserted main hallway. As I passed by the area containing the simultaneous interpretation booths, the door burst open and the lead interpreter emerged, visibly distressed: "Señor Martin, I tried to contact you to apologize. I am so sorry". Alarmed, I asked what was the problem?

"In my 30 years of international interpretation this has never happened. I apologize". Pressing her for details, I obviously needed to find a fix, quickly.

"Señor, while Jane spoke, we were unable to interpret, several times. We stopped; none of my team could talk because as they listened to Jane they began to cry, all of them. Even me!"

Smiling, I simply gave her a hug and said, "The power of commitment, love and hope is a solution, not a problem to be fixed, nor an incident needing an apology."

Afterwards, backstage, Jane beckoned me over and I told her the story. She said it was such a lovely story and it fit with what she wanted to ask me. "The atmosphere here is wonderful. I've spoken with Mary and we've shifted a few things. Would you mind if I stayed five days instead of two?" I smiled, again.

VANCE MARTIN – President Emeritus, Wilderness Foundation Global (South Africa). Founder/Co-Chair, Wilderness Specialist Group, IUCN/WCPA. WILD Foundation (1983-2023)

Dolphins and Dr. Goodall: Amazing Creatures and the Most Amazing Person

– William Archer –

My favorite days sailing are when I see dolphins. Their speed, power, grace, and joy – as they swim alongside and cross in front of the bow – always bring a smile. And my favorite days in general are the days when I speak with Jane. Her compassion, commitment and intelligence – as she tirelessly fights to save the planet and all living things on it, while simultaneously caring deeply about each of us who are privileged to know her – is immensely rejuvenating, just as her optimistic hope – despite fully understanding the greed, stupidity and ignorance that would destroy the planet – is profoundly inspiring. Dolphins are a joyful reminder of the magic of nature and Jane is an inspiring reminder of the best in humankind and what we can and must do to protect nature. It is not surprising, then, that one of Jane's many causes is to stop the cruel and unnecessary keeping and breeding of dolphins in captivity.

Many in this volume will undoubtedly and justifiably speak of Jane's boundless passion, compassion, energy, dedication, knowledge, intelligence, and leadership. I wish to highlight another trait well known by her family and friends. Jane is fun.

I recently told Jane that Chikako and I were planning to have a special dinner outside at a restaurant on a night that ended up having a heavy rain forecast. The day after the dinner, and after discussing other matters in an email, Jane asked how the dinner went and, if it rained, whether they rushed around with waterproof jackets and hoods, or umbrellas, and said she had a fun time imagining the two of us sitting with knife and fork in the air as our table floated away on a sudden rush of water! A few emails later, Jane wrote that an albatross brought her the news that Chikako and I were each riding a dolphin, and that we had managed to hang on to our glasses and had acquired a taste for brine! A few days later, and after the couple ended up on a sailboat, the story ended with two dolphins leaping out of the water and two transparent human-like shapes flying through the air, landing on the boat and seeming to vanish inside the man and woman.

We are all connected, and Jane reminds us of this in many ways. She is a living refutation of the cliched copout that "one person cannot change the world". And, let there be no doubt, Jane is also fun. Happy birthday, Jane – and cheers!

A Tribute to Dr. Jane Goodall

– Yvon Chouinard –

I've been influenced by Jane's work for at least 50 years.

At first, I was not particularly interested in chimpanzees, but more in Jane herself because she was the first behavioral scientist to discover that animals have intelligence and behaviors usually ascribed only to humans.

Because Jane lived with chimpanzees, and spent years in careful observation, she saw that they had relationships, built families, used tools, and even seemed to feel emotions. These were thought to be controversial findings because humans were thought to be entirely separate from animals, the only "tool makers."

Scientists today have mostly replaced years in the field with math and computers, but Jane's field work stood the test of time and is finally widely admired.

For me, the great importance of Jane's work is that she made the connection [or bridged the divide?] between us and other animals. If we are to save this planet, we must all develop this deep spiritual affinity with our natural world.

Because we are responsible for the mass extinction of many other creatures, beginning with the large mammals, we must acknowledge that we are a large mammal ourselves.

We do not have a God given right to be superior or more deserving of existence than the chimpanzee, our closest living relative. As we destroy nature, we destroy ourselves.

Jane has always faced hardships from living in the jungles of Gombe to challenges from the scientific community with steadfast courage. She was and remains entirely unconventional. Jane's selfless dedication, her steadfast beliefs, wisdom, and her soft -spoken way have in many ways inspired and influenced my own life and path in trying to save our home planet.

YVON CHOUINARD – Environmentalist and Founder of Patagonia, Inc.

Persistence in Kindness - Jane's travel companions spreading a message of hope and love. Drawn by Patrick McDonnell for the MUTTS comic strip. (mutts.com)

Image Source: Patrick McDonnell

Persistence

— Zara Bending —

Every time I experience 'Jane' live, it's never lost on me how she who wrote *In the Shadow of Man* lights up a room like no other. Some rooms are classrooms. Some are lecture halls. Then there are the theatres, arenas, Houses of Parliament, international forums, and even the United Nations General Assembly Hall. The best 'room' to play is Nature, of course. Anywhere outside with natural light and fresh air.

But then the clock struck 2020 and the only room was Zoom.

While a conservationist's natural habitat may involve long periods sat at a screen, we're definitely a migratory species that needs Nature to sustain us. We're also social animals so the lack of physical presence was harmful. The public learned a new word: 'zoonoses'. It seemed that our work across the network was more important than ever, but the looming uncertainty was debilitating, particularly for the young people at the heart of our work through Roots & Shoots; many of whom were already wracked with anxiety over climate change, mass extinction, and every other existential crisis bearing down on humankind.

From my screen in Sydney, I saw social media posts selling COVID-19 'cures' made from wildlife parts and derivatives. Had we learned nothing? I'd been a wildlife crime expert with JGI Global for a few years and the frustration I felt reminded me of Dr Goodall's talks about adversity and 'the indomitable human spirit'.

In the years that followed, I saw the indomitable human spirit as we persisted in our work to fight wildlife crime between lockdowns, waves, online conferences, podcasts, emails, and 'track changes'. I was proud to watch Dr

A live concert during the première of 'Jane' at the Hollywood Bowl in Los Angeles, in attendance by 17,000 people in 2017.

Image Source: Thomas D. Mangelsen

Goodall give a keynote speech to the 31st meeting of the INTERPOL Wildlife Crime Working Group in November 2020. The 'room' may have been in 'gallery view' but she still moved it and made an impact.

As a conservationist persisting in a time of grave uncertainty, Jane taught me that we can't surrender to circumstances beyond our control. This time it was a pandemic, but the same could apply to projects not going ahead, grants falling through, or changes in policy. In the absence of certainty, we should embrace consistency. The good news is that consistency in our actions and motivations is something we can choose against all odds. Choose to keep going every day, and if we all persist, we may have a chance at a future worth hoping for.

ZARA BENDING – Jane Goodall Institute Australia / Jane Goodall Institute Global (wildlife crime expert)

The Best Role Model in the World

– Zoe Weil –

Dear Jane,

When I was a child, if someone asked me what I wanted to be when I grew up, I would say, "I want to be Jane Goodall." I suppose that doesn't make me different from thousands of other girls growing up in the 60s and 70s who loved animals, were glued to National Geographic specials, and adored you. After a circuitous journey as a pre-med student, then an English major, and switching gears to (briefly) go to law school, I found my way back to my childhood dream. I read the book *The Mind of an Ape* and reached out to the author to work with the chimpanzees whom he was teaching a symbolic language.

I was so excited to meet the famous Sarah who had learned this language, but when I did, she screamed and bounced off the walls of her enclosure. It was so distressing to hear that she was no longer "complying" with the research and was permanently isolated, and to be told to stay away from the bars because she could rip my arm off if I got too close. But the next time I visited Sarah she was calm and came right over to me. I ignored the warnings and said to her, "Turn around and I'll scratch your back" as I spun my finger in the air. Sure enough, Sarah turned around, sank down to sit, and pressed her back against the bars. I scratched her back through those bars feeling heartbroken about Sarah's miserable life, so far from her jungle home in Africa, and I thought to myself, "What would Jane think of this?" I stopped working in that lab.

As much as I wanted to work with animals, I came to realize that I needed to work for them, and so I became a humane educator engaging hearts and minds in creating a more compassionate world. Then, a couple years after I launched a humane education program in 1989, I learned you had embraced humane education yourself and had created Roots & Shoots. I was following in your footsteps after all!

For decades I've taught young people about animal, environmental, and human rights issues; trained educators in integrating these issues into classrooms; spoken around the world; and written a bunch of books, and every single day that I've been a humane educator you've been my role model. When I sometimes feel despondent about the state of the world, I have your never-ending dedication and seemingly boundless energy to remind me that there is no giving up this vision of a humane future and pursuing it with full commitment.

Happy 90th birthday, Jane. Thank you for all you have done and continue to do to build a world where all life can thrive. I have been blessed beyond measure to have you as the best role model anyone could ever hope for.

With so much love and gratitude,
Zoe

ZOE WEIL – Co-founder and president of the Institute for Humane Education, which offers graduate programs in humane education, training for teachers and changemakers, and free downloadable resources. She is the author of award-winning books about building a humane, sustainable, and equitable world for all.

Afterword

We wondered how to close this book of out-standing tributes to an outstanding woman, and when Tom Mangelsen sent in his essay, we decided he actually did our work for us. Tom captured much of what others wrote and added some more wonderful personal and other stories.

After All Is Said and Done:
A Loving Birthday Tribute to Jane Goodall

— Thomas D. Mangelsen —

Dr. Jane Goodall, there are not enough adjectives in the dictionaries or on Google to describe your 90 years! Yes, you have inspired and changed millions of lives for the good around the globe, and you know you have indeed changed mine as well. You have brought awareness about so many species and individuals and saved so many lives through your films, books, lectures, writings, interviews and endless zooms. And it remains a mystery, beyond your great genes, just how you have been able to do so much for so many years. Your 300 plus days of world travel, your endless energy and passion to help save the earth from human's self destruction. These are traits millions know, Jane Goodall, the planet's animal rock star!

However, how many know the "other" Jane, the person I sometimes refer to as the "Real Jane"? Well, many writing tributes here know the Real Jane and I have been so blessed to be included in this group.

Like most, we began as "fans" of Jane and someway or another, were chosen to be FoJs or Friends of Jane, by Jane. Besides meeting her at a couple of book signings where our interaction was similar to many, going something like this, "Hi Jane, thanks so much for doing so much for the planet, would you mind signing your book for me?" I was asked to introduce Jane at a lecture in the spring of 2002 in Jackson Hole, which led to an 18 hour day trip together in Yellowstone Park. When I first met Jane backstage before her lecture we talked about our plan to go to Yellowstone and I realized I needn't be so nervous introducing her. She told me to make it short and added, "I

Jane and Tom during a picnic when exploring the Grand Teton
and Yellowstone National Parks, Wyoming, in 2002.

Image Source: Thomas D. Mangelsen

can tell my own story." Which I knew meant many introducers go on and on
about her incredible life. So, being a fan of her late husband filmmaker and
photographer Hugo van Lawick, I talked mostly about him!

Two days later, I picked up Jane to go off on our Yellowstone adventure.
"What would you like to see?" I asked. "Everything," she said. "Wolves are
five or six hours away, it will be a very long day," I replied. She assured me,
"As long as you get me to my flight at 9:30 in the morning we're good to
go." Wow, I thought, this is Jane Goodall. A free spirit wanting a break from
travel, people and cities and likely feeling she may never come this way again.

After driving an hour or so we stopped, Jane wanted to take a walk along
the shoreline. She picked up feathers and queried who they had belonged
to. She searched for the perfect flat stones and skipped them across the calm
waters of Yellowstone Lake, she was child-like and delighted in my lame at-
tempts, surely she had picked better stones! Early afternoon we had a picnic
lunch with French bread, cheese, veggies, dip and a nice bottle of sauvignon
blanc in a grassy meadow on the shore of the Yellowstone River. We talked

Jane observing cranes in Nebraska with Loup, Tom's dog, 2009

Image Source: Thomas D. Mangelsen

about all kinds of issues, personal and other things including cougars and cranes. We then carried on to the Lamar Valley where we saw wolves and a second grizzly and a family of black bears. I had never seen all three species in one day. "Jane Magic," she said. Driving home in the rainy night we finished the evening after midnight at her host's house with a whiskey and a video about cougars that I had been working on with my co-founder, Cara Blessley Lowe, and our newly formed nonprofit, The Cougar Fund.

Jane was sickened by the sight of hounds chasing and treeing a beautiful cougar, then shot and killed by a hunter! Jane immediately said, "I want to help with this cause, although my people always hate it when I take on another cause that is not related to my work with chimpanzees." However, Jane was not about to leave any animal cruelty issue pass if she could help, and wasn't about to be pigeonholed as only the "chimpanzee woman".

During our picnic on the Yellowstone River, Jane asked if I knew anything about the great crane migration in Nebraska? I told her of my film work with cranes over many years and having a family cabin on the Platte

River, in the heart of the migration. (Ironically, I realized later, that my NGS film, *Flight of the Whooping Crane* had competed for an Emmy in 1985 with her NGS *Among the Wild Chimpanzees*. Of course, her film won!). I asked, "What are you doing in March?" She said, "I'm coming with you to see the cranes!"

The following spring and for nearly 20 more years we have met at my cabin to witness the great sandhill crane and waterfowl migration. That first year while walking along the bank of the river I turned back and saw Jane with an armload of river trash. She had picked up beer cans, bottles and plastics. Jane walks the talk. Afterwards, she went back down the river bank to gather kindling and wood to make a fire. It was a chilly evening. I've learned over the years, how important a nice fire is to Jane, and, the colder and blustier outside, the better.

In March, of 2003, we were at the cabin. It was the evening of the beginning of the bombing in Iraq. We witnessed a most beautiful sunset in a hide as thousands of cranes, ducks and geese were landing against the setting sun. A half dozen deer ran through the flock, the cranes danced as the deer splashed between them. A bald eagle landed on the far sandbar. When we got back to the cabin we poured a whiskey, made a fire and talked of the wondrous evening. We had momentarily forgotten about the outside world.

In the morning we learned that bombs had been falling all night in Iraq. On the radio we heard about the vast destruction, the war was at full rage. We both said aloud, "How could this be happening after the previous evening?" Drinking our coffees we fell into a silence as Jane sat quietly in the corner of the kitchen peering out the window toward the field of corn stubble to the south. She then yelled, "Tom come quick, bring your camera, I want you to take this picture for me."

There, in the rising fog and drizzling rain, were hundreds of cranes dancing in the field as snow geese passed low over the cranes. I had taken the photo through the open window and thought it was a pretty scene but nothing to get too excited about. However, Jane, with all of her hope and wisdom over the years, saw more, she saw peace in the scene. Cranes are the symbol of peace around the world and she believed there would be peace again and that they were telling her so. She asked if I had captured it. I confirmed I had

The Peace Dance, named by Jane

Image Source: Thomas D. Mangelsen

and she named the photo and, later, the print "Peace Dance". Her signature, made with a silver Sharpie, remains on the window where she watched the cranes.

Most years Jane would invite good friends to join in for the crane gathering including Mary Lewis, her long time executive assistant, and Dave Matthews the famous musician. We picked Dave up at the nearby Kearney airport. He was wearing a hoody carrying only a gym bag and guitar. Soon after leaving the airport and seeing thousands of snow geese lifting off in the fields Dave exclaimed how his dad and his sister would love the sight! He was so genuine and funny we immediately connected.

After settling Dave into the one-room schoolhouse we had a fire, whiskey, dinner, wine and planned our early morning outing to see the cranes coming off their night time roosts. At dawn, I went to the river bank where I saw Dave in the fog slowly walking down the path to the river holding his iPhone high above his head. It seemed a bit odd at first, but of course Dave was capturing the ancient calls of cranes flying close overhead! The next afternoon, after a walk with the dogs, I saw Dave and Jane sitting and giggling in the old clawfoot bathtub in the backyard, much like my three brothers and I did when we were children. In those years, we had to take turns pumping the hand pump to fill the tub, then wait a couple of days for the water to get warm in the summer sun. First, dad took his bath, then the four boys and, lastly, mom. She was, indeed, the best mom ever! Now, 70 years later, in front of me was a sight I could have never imagined, Dr. Jane Goodall and Dave Matthews in that same bathtub! Two renegades, always stepping out of their "expected" roles.

The following evening after observing the sunset and cranes coming in to roost, Dave brought his guitar in and began playing, singing and serenading Jane who was sitting on the hearth of the fireplace. She had the most adoring look as she listened to Dave, and I could tell Dave was smitten. I videotaped the two of them with my small handycam from across the room. Although, it wasn't the best audio and video, it was the most precious moment ever in the cabin! I asked Dave later if I could post it on my website and he said, "Of course. If any of my agents call you, tell them to call me!" Dave,

Dave Matthews playing the guitar for Jane and others
gathered in Tom's cabin in Nebraska, 2011.

Image Source: Thomas D. Mangelsen

like Jane, isn't worried about the small stuff or being "managed". Jane had once told me that she was National Geographic's cover girl, of course she was, with her beauty and long legs. She also said, "When I was young I used to be a terrible flirt." Watching her and Dave, I couldn't help but think, you still have it girl!

In fact, I've seen her charm farmers in Nebraska that she wanted to have meetings with (she never takes a vacation, always trying to spread a conservation message) to talk about water issues and the amount of water taken out of the Platte to irrigate crops.

When she met with them farmers they greeted her with crossed-arms, wearing overalls and cowboy hats, tooth picks in the corner of their mouths. The half dozen men sat and waited for Jane to speak. It was obvious they were ready to tell this British woman that she had no business telling them how to run their farms. However, an hour later, after hearing Jane speak, they wanted Jane's autograph and to have their photos taken with her!

In South Dakota, when watching prairie dogs, she met with ranchers who hated the critters, poisoned them and shot them. Not only cruel but also limiting the recovery of black-footed ferrets, Jane did her magic with charm, poise and common sense. It was the same story as with the Nebraska farmers. Smiling ranchers left thinking a bit differently after Jane spoke with them, many kept Jane's autograph on slips of paper on their dash boards.

In the spring of 2006 Jane and I met in Rapid City. We were going to an event in Spearfish, South Dakota to have a meeting with the elders of the Sioux nation about Roots & Shoots becoming part of their teaching curriculum in the native schools. We had decided to meet a few days early in hopes of seeing the most endangered, once thought extinct, black-footed ferret. A friend of mine who had studied the ferrets for a number of years had invited us to take a small detour. For three nights we left the small town of Wall late in the evening in search of the nocturnal weasels, whose main prey are prairie dogs. For hours we drove many miles across the prairie looking with huge spotlights for the rarest of mammals. When we saw the reflection and unmistakable big green eyes of the first ferret on a hunt, running between prairie dog holes, we could barely contain ourselves. For two nights we searched and saw several more ferrets. Nights to remember!

Jane pretending to be a prairie dog doing a jumping-yipping display in South Dakota.
Image Source: Thomas D. Mangelsen

On our last afternoon we decided to have a picnic on a hillside overlooking a vast prairie dog town. We toasted the survival of the black-footed ferret, the prairie dogs, bison, pronghorn, birds and, as always, the "Cloud Contingent", those special souls that had passed before us, including our four-legged friends. The prairie dogs scurried from hole to hole to avoid predators. A red-tailed hawk high above set off a chorus of warning calls from the prairie dogs as they leaped in the air, barking to others to warn of the avian predator. Jane decided to try to help out and leapt in the air herself with her hands spread wide warning the entire dog town of the red-tailed hawk.

After the "all clear" sign was given by the chief dog we carried on to Spearfish. Dark clouds had been forming on the horizon. I had decided to take the more scenic country roads that we both preferred. It began to drizzle which soon turned to rain. The temperature was plummeting. Soon it was sleet, then big wet snow flakes. The dirt road became slippery and the snow was starting to stick. Although I had a Landcruiser I knew how important the event was and after a short chat with Jane, and common sense took over, we retreated to the nearest highway. Arriving at the Spearfish Lodge late afternoon, the snow was starting to accumulate. The historic log lodge with its giant river stone fireplace and roaring fire was a welcome site! The lodge had been reserved for Jane's event. A half dozen native Americans and a handful of Jane's friends had already arrived. The winds had picked up and it was doubtful others would make it.

By morning it was a full on prairie spring blizzard. No one was able to get in or out, all roads and airports were closed. Even the snow plow got stuck a mile away trying to open the road to the lodge. Visibility was near zero in the white out, Jane saw the shadow of a man walking and stumbling towards the lodge in waist deep snow. I put on my coat and boots and waddled through the drifting snow to help. The snow plow driver had abandoned his vehicle in hopes of getting to the lodge and safety. He was nearly frozen and hypothermic. Warmed with blankets, dry clothes, the fire and a whiskey he rallied.

Making the best of any situation Jane was quite happy to be snowed in. There were no options! Commitments for lectures or whatever, would simply have to be rescheduled for another year. Jane undeterred carried on with the few native elders who did make it and talked about Roots & Shoots.

In the evening Jane suggested we all gather around the fire, tell stories and sing songs. It was great fun! Meanwhile, Jane and I chuckled watching Jason, who used to work in my Santa Barbara gallery, and Patricia, a young beautiful native American girl sitting on the hearth talking, staring into each other's eyes, obviously unaware and who couldn't care less about the rest of us singing and telling stories! Jane had invited Jason to the event and Patricia was one of the only younger native women invited by the elders. Jane had gone to Santa Barbara a year or so earlier and gave a lecture which Jason and I attended. Jason told me some weeks later that he had been very depressed and had nearly taken his own life. And at the last moment he saw a note from Jane on his dresser and reread it. He told me Jane's note made him change his mind. When Jane learned about the near tragic incident she reached out to Jason. Now he and Patricia were holding hands. A few months later they moved in together at the Pine Ridge reservation and started the first native American Roots & Shoots program.

Whether it's farmers, ranchers, native Americans or movie stars like Pierce Brosnan, Angelina Jolie or Marlon Brando, Jane's humor, intellect, poise and charm is infectious. After the first Cougar Fund fundraising event in Laguna Beach in 2002 we went out on a whale watching boat with Pierce Brosnan and a dozen other friends. We saw nothing for an hour or more. Then Jane lay down in the bow to watch. Within minutes, the ocean exploded with hundreds if not thousands of porpoising dolphins from under Jane and the bow and as far as we could see. Jane magic! Mother Nature obviously looks fondly upon Jane.

After the event, Mary Lewis, Jane and I drove to Mulholland Drive in Los Angeles. Marlon Brando had invited Jane to interview her, something he had longed to do.

Arriving at Marlon's house we were led into the living room by one of his staff. Marlon was standing, his white hair pulled back into a pony tail, wearing a smart smoking jacket, long blue scarf, black baggy silk pants, and slippers. He walked to Jane and greeted her with a bear hug saying, "Darling, it's so great to see you again." then added that he had just read her book, *Reason for Hope*, and really loved it.

Marlon invited us to sit on the couches facing each other as two giant dogs greeted and then leaped onto Jane, making her day. Wine and food were served and Marlon asked about what we had been doing in town. Jane explained that we had a Cougar Fund event in Laguna and gave Marlon a *Spirit of the Rockies* cougar book. As Marlon looked through the pages he told us how he had lots of deer in his yard and how cougars would sometimes jump the fence and get them. He seemed to be fine with the naturalness of it all.

Marlon talked freely about Hollywood, his son's death, and much more. After a while his cameraman began to set up on the patio. Marlon and Jane went out to do their interview while we continued to eat and drink. After they wrapped up, we took a few snapshots of Jane and Marlon. Marlon was obviously enjoying his time with Jane and really wanted us to stay but Jane and Mary had a flight to catch. As we said our goodbyes I could tell Marlon was sad, as if he were very lonely.

Walking to the driveway I turned around to see Marlon with his arms outstretched between the sliding glass doors. I reached for my camera but I couldn't raise it to take his photo, although I doubt he would have minded. In that moment I saw an image of Marlon, like Jesus on the cross and just waved goodbye. It likely would have been one of the last photos of Marlon. He passed on July 1st, 2004.

While in Salt Lake City in 2002 for the Winter Olympics, Jane, Angelina Jolie, and some friends, including myself, went for a walk in the bitter cold down the street from the hotel. I don't remember how we ended up on a rooftop but all of a sudden I saw Jane jogging around its perimeter, stopping to tell others that if we wanted to stay warm we should all join in, which we did. Jane gets cold easily but knows how to survive. It was a memorable sight to see Jane, Angelina and the rest jogging on the roof. However, it wasn't as silly as it looked I guess. On the way back to the hotel Jane and Angelina sat on a bench next to one of those bronzed old man characters and asked me to take their picture. Jane and Angelina became good friends.

Over the years, Jane and I traveled to numerous cities together for lec-

Marc, Jane, and Tom toasting at a meeting of the executive committee of the Cougar Fund at Tom's Nebraska cabin in 2012.

Image Source: Cara Blessley Lowe

Tom Mangelsen and Jane
Image Source: Thomas D. Mangelsen

tures, Jane's film openings, fundraisers, and to many schools where Jane gave talks to her favorite of all, children, about her Roots & Shoots program. I know how much energy all those events took, all day, days on end, every hour with an interview, lecture or meeting. I was worn out after a couple of days, yet she carried on to spread her message, she in her seventies and then eighties.

Unbelievably, she continues to maintain the same grueling schedule today. Some say it's even worse!

Besides Yellowstone and the cabin and "crane time", as we call it, the other most fun times with Jane were when we were at my home in Moose, Wyoming where she could relax a bit. She loved playing with and throwing balls in the pond for whatever dog or dogs I had at the time.

Most everyone knows Jane's fondness for dogs surpasses all animals, even chimps! At the cabin in Nebraska I was always sure to take a dog so she could get her "dog fix". Nothing made her happier. Jane loved riding around Teton Park in my red 1949 Studebaker truck with Loup, her and my favorite

Jane walking in Gombe in 2018.

Image Source: Thomas D. Mangelsen

Labrador, at her side. During the summer of 2010 Jane spent a month alone at my cabin writing one of her books with Loup as her only companion. I was in Brazil's Pantanal photographing jaguars. Jane told me later that those 30 or more days were the longest stretch of time she had been by herself in over twenty some years. I thought how great it was that she could have some serious quiet time and alone time throwing balls in the pond for Loup.

There are so many stories with Jane but the ones I love most are when she is out there in wild places among the animals and places she loves jumping with prairie dogs.

THOMAS D. MANGELSEN – Award-winning photographer

mangelsen.com

We are all connected - Jane kissing her 'Prince Frog'.
Please transform your life and that of others by connecting with a local Jane Goodall
Institute or Roots & Shoots group. Visit janegoodall.global and rootsandshoots.global

Image Source: Thomas Mangelsen

About the Editors

Marc Bekoff, Ph.D. is a professor emeritus of Ecology and Evolutionary Biology at the University of Colorado, Boulder. He has published 31 books (or 41 depending on you count multi-volume encyclopedias), won many awards for his research on animal behavior, animal emotions (cognitive ethology), compassionate conservation, and animal protection, works closely with Jane Goodall, and is co-chair of the ethics committee of the Jane Goodall Institute, and a former Guggenheim Fellow. He and Jane wrote *The Ten Trusts: What We Must Do to Care For the Animals We Love.* Marc also works closely with inmates at the Boulder County Jail as part of Roots & Shoots. In June 2022 Marc was recognized as a Hero by the Academy of Dog Trainers. His latest books include *Canine Confidential: Why Dogs Do What They Do, A Dog's World: Imagining the Lives of Dogs in a World Without Humans, Dogs Demystified: An A to Z Guide to All Things Canine*, and the second edition of *The Emotional Lives of Animals.* Marc also publishes regularly for *Psychology Today*. His homepage is marcbekoff.com. In 1986 Marc won the Master's Tour du Haut cycling race, aka the age-graded Tour de France.

KOEN MARGODT, PH.D. has been working with Jane and volunteering for the Jane Goodall Institute since 1995. Together with Marc Bekoff, he is Co-chair of the Global Ethics Committee of the Jane Goodall Institute, advising Jane and thirty JGI chapters on a wide variety of ethical topics. Together with Melody Horrill, he's also Co-chair of the Cetacean Committee of the Jane Goodall Institute, which seeks to phase out the keeping of dolphins in captivity. Several statements can be found on thejanegoodallinstitute.com/news. Koen did a Ph.D. in Moral Philosophy on the moral status of great apes at Ghent University, Belgium. He's the author of *The Welfare Ark: Suggestions for a Renewed Policy for Zoos* (2000), in which he makes a plea to turn zoos into sanctuaries. He's also a Guest Professor at various universities, where he lectures on animal ethics. As his "day job" that usually extends into evenings, he carries the hat of Senior IT Training Manager at an international law firm. He lives with his wife Iris, their daughters Mana and Fara, dog Lola, rescued cats Lotte and Quinn and four chickens in the green hills surrounding Leuven, Belgium.

koenmargodt.com

www.ingramcontent.com/pod-product-compliance
Ingram Content Group UK Ltd.
Pitfield, Milton Keynes, MK11 3LW, UK
UKHW050005301025
8653UKWH00075B/714